for Vanna

every person's guide to
antioxidants

JOHN R. SMYTHIES M.D. F.R.C.P.

RUTGERS UNIVERSITY PRESS
NEW BRUNSWICK, NEW JERSEY, AND LONDON

Library of Congress Cataloging-in-Publication Data

Smythies, John R. (John Raymond), 1922–
     Every person's guide to antioxidants / John R. Smythies.
          p.   cm.
     Includes bibliographical references and index.
     ISBN 0-8135-2574-8 (alk. paper). — ISBN 0-8135-2575-6 (pbk. : alk. paper)
     1. Oxidation, Physiological.   2. Antioxidants—Health aspects.   3. Free
radicals (Chemistry)—Pathophysiology.   I. Title.
     RB170.S69   1998
     616.07—dc21   98-6810
                                                                                   CIP

British Cataloging-in-Publication information available

Manufactured in the United States of America

every person's guide to
antioxidants

contents

I am most grateful to Lesley Smythies for her excellent and painstaking editorial work on this book and her advice about immunology; to Christopher Smythies for helpful comments from the point of view of neurosurgery and for supplying the epitaph from Tombstone; and to Vanna Smythies for help in adjusting the material for the lay reader. I should also like to thank my dedicated agent Julie Popkin; Charles Thomas of Pantox Laboratories, San Diego; Jessica Hornik Evans, copy-editor; and Doreen Valentine of Rutgers University Press for her skilled and tireless editorial guidance. I should also like to record the debt that we all owe to the Ireland family of Birmingham, Alabama, pioneers in the active support of promoting the importance of antioxidant vitamins in health and disease.

acknowledgments

every person's guide to
antioxidants

introduction

We live in an age in which millions of people in the United States and elsewhere take vitamins to supplement their diet under the belief that these will help to maintain good health and ward off illness. An enormous industry for manufacturing, packaging, and distributing vitamins has grown up to meet this need. In the United States there are more than eleven thousand health-food stores and nutrition centers that sell only vitamins and related products. Furthermore, nearly every drugstore and supermarket has an extensive area devoted to vitamin preparations, of which there are hundreds of varieties and brand names. These vitamins are sold either as mixtures of multivitamins and essential minerals or individually. Among the "special" types of vitamins that fill these shelves are those labeled "antioxidants." These include such familiar vitamins as A, C, E, and beta-carotene. Less well-known antioxidants include melatonin, lycopene, zeaxanthin, and various flavonoids.

Newspapers, magazines, and talk shows are brimming with discussions of antioxidants. These accounts claim that antioxidants are effective in helping to prevent cancer, heart disease, and other chronic illnesses. But the mass media also carry reports on some alleged alarming side effects of antioxidants. For example, the November 25, 1996, issue of *Time* magazine carried a long article called "Can We Stay Young?," which stated that, although some nutritionists have recommended a diet high in fruits and vegetables that contain antioxidants to combat disease, this approach has an "uneven record." According to the article, in some studies the use of antioxidants appears to be associated with a "dramatic" reduction in cancer and other diseases, but in other studies beta-carotene (the only antioxidant mentioned by name in the article) actually seems to be associated with an increase in cancer. The article concluded: "In either event few contemporary aging researchers think self-medicating at a salad bar is the best way to extend the human life span." This report is seriously misleading, as this book shows, and indicates the level of confusion that the public faces with regard to the science of antioxidants. The March 1997 issue of *Consumer Reports* carried a section on antioxidants that also provided consumers with conflicting information. The article correctly stated that vitamin E had been shown to be protective against heart attacks but cast doubt on the effectiveness of beta-carotene and vitamin C as protective agents. However, the article neglected to mention a most important fact: antioxidants should always be given as a well-balanced mixture (either in the diet or as supplements) and not singly. In its August 1997 issue, *Consumer Reports* had a section on methods that women should use to reduce their risk of a heart attack. It failed to mention any dietary factors other than a low fat and alcohol intake.

The National Academy of Sciences has for many years published a list of recommended daily allowances (RDAs) for the common vitamins. For example, the current RDA for vitamin C is 60 mg per day and for vitamin E, 20 mg per day. But frequently the dosages of vitamins per tablet on the market contain many times the amount of the recommended daily allowance. Why so? Some twenty-five years ago Nobel Laureate Linus Pauling claimed that people would benefit if they took "megadoses" of vitamins. His argument was that the recommended daily allowances reflected only what was needed to avoid specific vitamin deficiency diseases such as scurvy, pellagra, and beri-beri. What is really needed, he claimed, is "optimum" doses: if 10 mg of a vitamin is beneficial, then 100 mg is bound to be ten times better. Although there was little evidence to support this claim at the time, it soon became a popular belief and the basis for marketing megavitamins. Pauling was correct in his observation that the recommended daily allowances of vitamins today are calculated on the basis of the amount needed to avoid deficiency diseases, but he never gave any convincing reason for his idea of optimum levels of intake. Pauling's ideas were not received with enthusiasm by the medical establishment; in fact, the general opinion was that they amounted to quackery. We have now come to realize, on the basis of facts that were not known twenty-five years ago, that the amount of each antioxidant vitamin needed to avoid deficiency diseases is indeed not the same as the amount that a healthy body needs, as Dr. Leland Tolbert and I pointed out in a paper published in 1981 [193]. This is because antioxidant vitamins play a very important role in the body that is quite distinct from their particular role in preventing scurvy and other deficiency diseases: they provide the body with its own antioxidant defenses. It would be fitting to reevaluate the role of Linus Pauling in medicine. A famous epitaph on a gravestone in a Tombstone, Arizona, cemetery, sums up this situation well: "Here lies the body of George Thompson/ Hung for murder 1882/ He was right and we was wrong/ But we strung him up/ and now he's gone."

Although we've all heard the term "antioxidants," many people are not very clear about what antioxidants are, what they are supposed to do, and if—and when and why—it is advisable to take them. This book describes what antioxidants are and how they work. To help the reader understand the role played by antioxidants, I discuss oxidative stress, which is due to the overproduction of potentially harmful oxidants in the body. It is the job of antioxidants to counteract the deleterious effects of harmful levels of oxidants. Oxidants play many normal roles in the body; it is only their overproduction (or the failure of antioxidant defenses) that results in harmful oxidative stress (i.e., disease). Antioxidants may be taken either by healthy people in an attempt to ward off the development of chronic diseases like cancer and heart attacks, or by sick people who are facing a disease for which there is evidence that antioxidant therapy will help. I explore in depth the relationship between oxidants and oxidative stress and disease and look at how antioxidants may function to prevent or combat disease.

Many scientific experiments and clinical trials have been carried out to test the idea that antioxidants are important to maintain good health. In this book I review the current medical and scientific literature on oxidative stress and antioxidants and present the most important original data from these scientific experiments and clinical trials, with their good and bad points. It is my hope is that, armed with an understanding of the research and a fair evaluation of the results as presented here, the reader will be able to decide for him- or herself as to the need to take antioxidants. Such a decision will be based on informed knowledge of the facts rather than on the propaganda put out by uncritical enthusiasts either for or against taking antioxidant supplements. This information will also enable the reader to discuss antioxidants with a doctor. It is still unfortunately true that many physicians have not kept up with the recent advances in this field, which is still regarded by some as tainted with the "alternative medicine" stigma. For those in the health professions, this book will help establish that antioxidants are important for the care of patients.

My credentials for writing this book derive from many years of scientific research on oxidative stress and antioxidants. In particular, I have studied the role that oxidative stress plays in schizophrenia; I was the codeveloper in 1952 of the first specific biochemical theory of schizophrenia—the transmethylation hypothesis. In 1954 Dr. Abram Hoffer, Dr. Humphry Osmond, and I discovered that an oxidized derivative of epinephrine, one of the hormones that is secreted by the adrenal gland and is also found in the brain, produces psychotic symptoms in nonpsychotic volunteers [89]. Although my work has focused on the relationship between oxidative stress and psychiatric illness, I have also studied the role of oxidative stress and antioxidants in all diseases.

This book is directed at all people interested in the question of whether they need to change their diet in order to improve their long-term health and whether they also need to take antioxidant supplements to do so. It is also directed at health professionals involved in preventative medicine and in treating the diseases covered in the book. This book will also be of interest to those who work for government health programs, including Medicare, and nongovernment organizations concerned with healthcare delivery; the health insurance industry; and any others who are responsible for public health policy and funding. The measures this book supports have the potential not only to reduce the amount of chronic diseases suffered by people but also substantially to reduce the current crippling costs of medical care.

Knowledge of chemistry is not necessary to read this book.

the basics: oxygen, reactive

oxygen species, and oxidative stress

before we can talk clearly about what antioxidants are and how they prevent disease, we need to consider some basic concepts about the body's cells and organs. Every cell in the body—whether a heart cell, a liver cell, or a brain cell—functions like a miniature chemical factory. A large number of different types of chemical reactions occur within the cell, leading to the breakdown of large complex molecules into smaller products or to the synthesis of new molecules from smaller building blocks. Other chemical reactions occur that may lead to the transfer of small electrical charges from one chemical substance to another. Among these transfer reactions are oxidation—in which a negatively charged particle called an electron is lost—and reduction, in which the electron is gained. Burning is one form of oxidation, as when coal (carbon) burns to form carbon dioxide. But a more general form of oxidation involves this transfer of electrons from one molecule to another.

Curiously enough, oxidative stress arises inevitably from the chemistry of the life-giving molecule of oxygen. Except for

part 1

some very primitive bacteria, oxygen is essential for life and provides the energy on which all cells in the body operate. That is the good news. The bad news is that the ordinary oxygen molecule easily turns into oxidizing agents, called reactive oxygen species, that possess great potential danger. In simple terms oxidizing agents (also called pro-oxidants) are short of electrons (each atom has a nucleus, composed of protons and neutrons, surrounded by negatively charged electrons) and will steal them from any neighboring molecule in the body that does not keep a tight hold on its own electrons. This damages the neighboring molecule severely. Common household bleach is an example of a powerful oxidizing agent in which the active ingredient is the poisonous gas chlorine. Reactive oxygen species play normal roles in the body, but in excess they develop highly poisonous properties.

The most important reactive oxygen species in the body are the superoxide ion ($O\cdot^-$), the hydroxyl radical ($\cdot OH$), and hydrogen peroxide (H_2O_2). The stable molecule of water is made up of two atoms of hydrogen and one of oxygen (H_2O). Note that the highly toxic hydroxyl radical is composed of one atom of oxygen and only one of hydrogen. Hydrogen peroxide, used widely in households as an antiseptic, is made up of two atoms of hydrogen and two atoms of oxygen. These reactive oxygen species are all short of electrons and are thus powerful oxidizing agents. Hydrogen peroxide is actually somewhat different, as it is not itself short of electrons but easily converts in the body to compounds, such as the hydroxyl radical, that are short of electrons and that do the damage. Technically, compounds that are short of electrons are called free radicals. The superoxide ion and the hydroxyl radical are free radicals. Thus, reactive oxygen species include both oxygen-derived free radicals and compounds like hydrogen peroxide that are not themselves free radicals but that easily generate them. Left free to react in the body's tissues, reactive oxygen species will attack and damage key molecules in the body, such as fats, proteins, and DNA— the molecule that carries genetic information.

To understand why reactive oxygen species are potentially so harmful for cells, we have to look a bit more closely at the cell itself. Each cell in the body is composed of a membrane boundary, a fluid interior, and a central nucleus, along with a set of other small structures called organelles. The membrane of the cell is made largely of fat. When the fats in the membrane are oxidized by a reactive oxygen species, the membrane becomes brittle and leaky; eventually, it falls apart and the cell dies. This is a lot like butter turning rancid, which happens when the fats in butter get oxidized. Hence dairies put antioxidants in the butter to prevent this from happening. Proteins are located either embedded in the fatty membrane or throughout the interior of the cell. Many proteins are small machines that perform important functions in the cell. Other proteins have a purely structural role. Some proteins are enzymes that manufacture substances needed by the cell, and some are hormones that act as signals to other cells. Others control the entry and exit of substances across the cell membrane. When a reactive oxygen species attacks a protein and damages it, some key function of the cell will be jeopardized. DNA, the molecule that carries the genetic information of the cell, is found in the cell's nucleus. Oxidative damage to it can cause mutations that predispose the cell to cancer formation. In these ways, through damage to fats, proteins, and DNA, various parts of the body can be weakened by oxidative attack and will succumb to a wide variety of diseases. I explain this fully in part 2.

If oxidizing agents are so harmful to cells, we have to ask why nature would have produced a cellular system that manufactures them. Indeed, several essential biochemical operations in the body generate reactive oxygen species as a part of the essential mechanism by which they work. Take several examples:

1. The basic mechanism of energy production of the cell produces reactive oxygen species as an inescapable by-product of the necessary chemistry involved. It has been estimated that 10 percent of the oxygen we breathe is turned into reactive oxygen species during this process.

2. The job of some white blood cells of the body's immune system is to attack and kill invading pathogenic bacteria and viruses. One weapon they use to do this is highly poisonous reactive oxygen species.

3. During the process of inflammation a set of chemicals called prostaglandins is secreted. Prostaglandins cause the redness, pain, and swelling associated with inflammation. They are involved, for example, in infections, burns, and arthritis. A key enzyme (called prostaglandin H synthase, or PGH synthase for short) that makes them

generates large amounts of reactive oxygen species as a by-product. Aspirin works by switching off this enzyme. This results in less prostaglandin-induced pain and swelling.

4. As the brain develops, many more nerve cells and many more connections between them are made than are actually needed. As the brain grows during childhood, the excess numbers of these are pruned away. During this process reactive oxygen species molecules are probably used as a pruning agent to kill the unwanted cells. During the learning process in adults as well as in children, many new connections between nerves (called synapses) are made and old, failed, ones removed. Reactive oxygen species are probably involved in the selective removal of unwanted synapses (as is discussed later in more detail). Reactive oxygen species also have normal functions in relation to the control of DNA action. They do this by activating a molecule called NF-κB. This molecule switches on a number of genes, including those responsible for making key molecules concerned in inflammation. The present focus of attention of researchers is on the role of reactive oxygen species in the mechanisms by which cells signal to each other during the inflammation process [222]. The indirect function of reactive oxygen species operates in addition to the direct attack by reactive oxygen species on proteins, fats, and DNA. This new research is important, as abnormalities in NF-κB activation are involved in atherosclerosis, Alzheimer's disease, HIV infection, rheumatoid arthritis, asthma, and other disorders.

These processes illustrate the fact that reactive oxygen species are not always villains but have some normal functions in the body. It is only when there is some disturbance in this system that disease results. Simonian and Coyle [185] give a graphic account of oxidative stress as "an expanding, self-perpetuating, and reinforcing series of metabolic events, which promote the generation of [more] reactive oxygen species and impair potential protective mechanisms. Like a spreading wildfire, the site of the initiating spark may be obscured in its terminal stages."

Oxidative stress can result from four main causes: (1) Reactive oxygen species can be ingested in excess from some environmental source (such as tobacco smoke or diesel oil fumes); (2) the diet may contain insufficient antioxidants; (3) there may be some disturbance in the elaborate biochemical systems that control their production and distribution (as in the case of many diseases); (4) there may be a failure in protective antioxidant mechanisms (as in the case of many other diseases). The topic of oxidative stress is dealt with in detail in part 2.

antioxidants The biochemical reactions that produce reactive oxygen species are essential for life and can be traced very early on in evolution. They are found in cells ranging from bacteria to human cells. But, as we discovered above, these essential, naturally occurring substances can also damage and kill cells. So how is it that cells survive despite the constant onslaught of reactive oxygen species? The answer is that evolution had to produce efficient defenses that would prevent reactive oxygen species from killing the cells. These defenses consist of a set of specialized molecules known as antioxidants. As their name suggests, antioxidants function to prevent the damage to cells that would otherwise occur as a result of an attack by reactive oxygen species.

To understand how antioxidants perform their role and why different types are needed, we must return to the structure of the cell itself. The interior of the cell and the fluid between cells are composed mainly of water. The cell membrane, as we have seen, is made largely of fat. As we know from experience, oil and water do not mix. Chemically speaking, the general rule is that substances that are soluble in water are not soluble in fat, and vice versa.

As oxidants can strike either at the fatty cell membrane or at the watery cell contents, antioxidants are needed that will work either in the fatty membrane or in the watery compartments. In other words, the cell needs to have available antioxidants that are soluble in water and others that are soluble in fat. Besides solubility, another way in which antioxidants differ from each other is size. Some antioxidants are small molecules, whereas others are large proteins. The small molecule types work in part by mopping up or "scavenging" the reactive oxygen species and carrying them away, and in part by neutralizing them chemically. The protein antioxidants are either enzymes that turn reactive oxygen species into harmless substances, or they are inessential "sacrificial" proteins (like albumin). A sacrificial protein will absorb reactive oxygen species and thus prevent them from attacking some essential protein.

The body gets its essential antioxidants from two main sources: some antioxidants are produced in the body itself, whereas others have to be obtained from the diet. Some of the dietary antioxidants are called vitamins—meaning "vital amine"—because they are essential for life, and low levels of them in the diet cause deficiency diseases such as scurvy and pellagra. There are other dietary antioxidants that the body cannot synthesize, but these are not essential for life, and their absence in the diet does not lead to disease. For this reason these antioxidants are not called vitamins. Although they are not essential ingredients in the diet, they are nevertheless helpful in combating oxidative stress.

SMALL-MOLECULE ANTIOXIDANTS

Of the small-molecule types the chief water-soluble antioxidants are vitamin C and glutathione. Vitamin C, whose chemical name is ascorbic acid or ascorbate, is a relative of the simple sugar glucose. Glutathione, a less familiar although just as important antioxidant, is a small, proteinlike molecule. Humans cannot make their own vitamin C, so we must acquire it from our diet; but most animals can produce their own vitamin C. All animals, including humans, can make glutathione. When natural levels of glutathione are low, N-acetyl cysteine (NAC) can be administered clinically as an antioxidant. A precursor molecule, NAC is turned into glutathione in the body. In the brain, vitamin C is the main antioxidant in the fluid between nerve cells, and glutathione is the main antioxidant inside them.

The main fat-soluble, small-molecule antioxidants are vitamin E, various carotenes (relatives of vitamin A), lipoic acid, and vitamin Q_{10}. Vitamin E is alpha-tocopherol. It has a relative—gamma-tocopherol—that is also a significant antioxidant. Vitamin E also boosts immune responses, both by protecting the membranes of the rapidly dividing immune cells in the tissues from oxidative stress and by protecting the sympathetic nerves in the lymphoid tissue, where the immune cells are formed, against oxidative stress.

The carotenes are the red and yellow pigments found in red and yellow vegetables; the best known is beta-carotene. However, only 25 percent of the carotenes in the diet consist of beta-carotene. Beta-carotene is particularly important in the ovary, where it is found in very high concentration in the corpus luteum, the ovarian structure that sustains pregnancy by making the key pregnancy hormone progesterone. There are over five hundred other carotenes found in fruits and vegetables in the diet, the most important of which are lycopene, reservatol, zeaxanthin, and lutein. Lipoic acid regenerates the active forms of vitamins C and E. It is found in dark green leafy vegetables.

Small-molecule, fat-soluble antioxidants have to be obtained from plants because animals, including humans, cannot make them. In the case of this class of antioxidants, there are significant chemical differences between the natural forms found in plants and the forms synthesized by chemists. The natural form consists of one stereoisomer, the synthetic variety of eight. The various stereoisomers have the same chemical formula but are put together differently and may have different biological activities. Vitamin Q_{10} is found in meat, with high levels in organ meats such as liver and kidney. Other important antioxidants in the body, particularly in the blood, are uric acid and bilirubin; melatonin, a powerful antioxidant produced by the pineal gland, is involved in the regulation of sleep.

LARGE-MOLECULE

PROTEIN ANTIOXIDANTS   Some of the large-molecule antioxidants are enzymes that turn reactive oxygen species into inactive products. The main ones are superoxide dismutase (SOD), which detoxifies the superoxide ion ($O\cdot-$); catalase (CAT), which deals with hydrogen peroxide (H_2O_2); and glutathione peroxidase (GSHpx), which takes care of several types of reactive oxygen species. These are all essential enzymes and are synthesized in the body's cells. Other large-molecule antioxidants are large inessential proteins (like serum albumin) that mop up reactive oxygen species and get damaged in the process. Their "sacrificial" function protects essential proteins, like enzymes and DNA, by getting in the way of the bullet, as it were.

SYNTHETIC MOLECULES   The drug industry is busy making new synthetic antioxidants that may be more effective than the natural ones in combating disease. These include steroids (such as the lazaroid group), fat-soluble derivatives of vitamin C, mifepristone, and many others, some of which are mentioned later in this book.

SYNERGISM BETWEEN
ANTIOXIDANTS   Vitamin C acts both on its own and in cooperation with vitamin E. It does the latter because, when vitamin E has mopped up a reactive oxygen species, the vitamin E is itself oxidized. And, to be effective, this oxidized vitamin E must be turned back to the active form; this is what vitamin C does. Once the vitamin C has been oxidized, it too must be turned back to its active form. This is done by another compound that includes a form of vitamin B (nicotinamide) in its molecule. Thus, the nicotinamide form of vitamin B also has indirect antioxidant properties. This chainlike helping or synergistic process has important consequences.

Carotenoids are also involved in recycling vitamin E in another chain: vitamin E-carotenoid-vitamin C. However, vitamin C has important antioxidant functions on its own in addition to its connection with recycling vitamin E.

IMPORTANT ANTIOXIDANTS
FOUND IN PLANTS   Flavonoids and phenols (which together are also called phytochemicals) form a large class of natural antioxidants present in many plants, with more than six hundred antioxidant phytochemicals having been detected. Rich sources of these agents are tea, garlic, olive oil, and many herbs, fruits, and vegetables. It has been estimated that the average daily intake of total flavonoids in the United States is about 500 mg. They act in several ways—by mopping up reactive oxygen species and toxic heavy metals, by preventing the formation of reactive oxygen species and fat oxidation, and by protecting other antioxidants such as vitamin C. They also inhibit LDL (bad cholesterol) oxidation, thereby sparing the precious vitamin E. However, once all the vitamin

E has been consumed by an oxidative attack, flavonoids fail to inhibit fat oxidation. This suggests that their action is dependent on the presence of vitamin E—that is, they belong to the group of vitamin E "helpers." But they act in many other ways besides being antioxidants. Many flavonoids interact with cytokines (key proteins in the cell that act as messengers, particularly in connection with inflammation), and they also block the production of inflammatory prostaglandins and prevent blood platelets from sticking together to form clots. 1/5

Antioxidant flavonoids include catechin (green tea), gossypol (rice), apigenin (chamomile tea), quercetin (apples), hesperetin (oranges), and naringenin (grapefruit). Antioxidant phenols include thymol and carvacuol (thyme), ferulic acid (many herbs), gallic acid (nutgall), hydroxytyrosol (olive oil), fisetin, morin, and many others. Antioxidant polyphenols are found in oranges (as cryptoxanthin) and many other fruits, tea, coffee, chocolate, licorice, and white wine. Other related plant antioxidants include rhein (rhubarb) and aloe-emodin (aloe). Antioxidant peptides (small, proteinlike molecules) include carnosine (which is particularly associated with the glutamate receptor in the brain) and anserine. Rosemary contains compounds that stimulate the production of the antioxidant glutathione in the liver and act as antitumor compounds that block cancer-producing chemicals from binding to DNA.

SOY BEANS. Soy beans contain a variety of anticancer compounds, such as genistin, which are effective at mopping up reactive oxygen species but are also likely to act in additional ways to kill cancer cells. Genistin and related compounds are also found in whole grain cereals, seeds, berries, and nuts.

RED AND WHITE WINE. Red wine contains up to twenty times more flavonoids and related compounds than does white wine and ten times more than tea. These include quercetin, rutin, and catechin. Red wine also contains a potent antioxidant—reservatol—found also in peanuts. Reservatol slows the making of fats by the liver and inhibits prostaglandin synthesis (and thus inflammation); it also prevents platelets from sticking together (a major cause of the blockage of coronary arteries leading to heart attacks). A derivative of reservatol is an ingredient in a folk medicine called kojo-kon. It is also found in another folk medicine obtained from *polygonum cuspidatum*, which has been used since ancient times for the treatment of heart disease, allergies, and inflammation. It has been suggested that some of the reported beneficial effects of red wine may actually be attributable to raised blood levels of the major blood antioxidant uric acid. Port wine raises blood urate levels, but an equal amount of pure alcohol does not. Port drinkers tend to get gout, which stems from a disorder of uric acid metabolism.

TEA. The catechins and related flavins in tea have been reported to inhibit fat oxidation in red cell membranes and protect DNA against hydrogen peroxide. A tea antioxidant—epigallocatechin gallate—inhibits oxidative damage to DNA. A test-tube experiment showed that flavonoids from green tea protected artificial fat bilayer membranes from fat oxidation. In the case of alcoholic extracts, green tea is much more potent than black tea [224]. Therefore, some other unknown antioxidant compounds must be present in green tea that are soluble in alcohol but not in water. Watery solutions of green tea and black tea (which both contain quercetin, myrecetin, and kaempferol) are equally effective in preventing fat oxidation in test tube experiments. These compounds also have anticancer effects in a number of tests.

GARLIC. In Mediterranean countries garlic has always been regarded as a health-giving food. There is a saying in Italy that to live long one must eat plenty of *aglio, olio, e limone* (garlic, olive oil, and lemons). The health-giving effect of garlic is not just folk wisdom; garlic contains the potent antioxidants alliin and allicin, which are sulphur-containing amino acids.

OLIVE OIL. Olive oil contains a number of antioxidant compounds (including beta-tyrosol, hydroxytyrosol, caffeic acid, and vanillic acid); the concentration of these substances is higher in extra virgin oil than in refined products. The bitter principle of olive oil (oleuropein) also protects LDLs against fat oxidation. A potent phenolic antioxidant in olive oil has the effect of preventing blood platelets from sticking together and reduces the production of inflammation-producing thromboxanes (relatives of prostaglandins). Interestingly, it has now been discovered that the waste waters from olive oil production are particularly rich in a number of antioxidants.

OTHER PLANT ANTIOXIDANTS. Turmeric, an ingredient of curry powder, contains the antioxidant curcumin, a polyphenol. In animal experiments long-term feeding of curcumin significantly reduced the ability of carcinogens to produce colon cancer. Yeast and yeast extracts (such as Marmite) are loaded with antioxidants (including glutathione and vitamin Q_{10}).

Many herbal medicines have a high content of antioxidant phytochemicals. An ayurvedic—the ancient folk medicine of India—drug prepared from a number of plants is marketed today as a product named Geniforte. In animal experiments this induces the synthesis of antioxidant enzymes and reduces the level of lipid oxidation. Boldine, an ingredient in a Chilean folk medicine made from the boldo tree, is used for the treatment of liver complaints and rheumatism. Very potent at mopping up hydroxyl radicals, this antioxidant is nontoxic and nonmutagenic.

It is likely that many herbal remedies work by virtue of their antioxidant content, yet the reverse is also true. Some flavonoids have properties that depend on actions unconnected with their antioxidant properties. For example, it has been shown that some flavonoids, like quercetin, gossypol, chrysin, and apigenin, bind to a certain receptor in the brain to which benzodiazepine tranquilizers, such as Valium, usually bind. The herbs from which these compounds are obtained (such as *passiflora coerulea* and *matricaria recutitis*) have been reported in folk medicine to have sedative and pain-relieving properties. A synthetic derivative from these flavonoids has been produced that is one hundred times as potent as Valium [137]. Moreover, it lacked some of the annoying side effects of Valium, such as sedation and muscle relaxation.

isoprenoids   These are not antioxidants but anticancer agents that retard tumor growth by inhibiting cell proliferation. Examples of isoprenoids are alpha-limonene (found in citrus oils), geraniol, carvone, and menthol. Rich sources in the diet are herbs, spices, barley, rice bran, olives, wine, eggs, dairy products, and certain essential oils used in flavoring. Isoprenoids in animal experiments also lead to the increased production of a group of enzymes that detoxify many environmental poisons. Isoprenoids also increase the liver output of two enzymes concerned in antioxidant defenses. Thus, isoprenoids have indirect antioxidant properties. Not all beneficial agents in fruit and vegetables are antioxidants. If you decide to rely on antioxidant supplements rather than increasing the fruits and vegetables in your diet, you will be depriving yourself of the added protection offered by isoprenoids (and probably other as yet unknown chemicals) against cancer and many environmental poisons. It is to be hoped that some isoprenoids may be added to antioxidant supplements.

the role of oxidative stress and

antioxidants in health and disease

In part 1 we saw how the normal chemistry of oxygen inevitably leads to the production in the body of reactive oxygen species, which play a normal role in important bodily functions. However, these agents are potentially toxic, and evolution has provided the body with a series of biochemical defenses against them. In many diseases the delicate balance between reactive oxygen species and antioxidants is disturbed. If reactive oxygen species are produced at a rate higher than what the antioxidant defenses can cope with, oxidative stress results, which may lead to tissue damage and disease. Medical science is exploring the possibility of using antioxidants in the prevention and cure of disease.

The megavitamin concept was once regarded with great suspicion by practitioners of mainstream medicine. According to Rucker and Stites, writing in 1994, "a decade ago, it was considered highly speculative to propose approaches to disease prevention that involved supplementation. Nutrition paradigms important to vitamin function, however, have now shifted to include more focus on disease prevention and consideration

part 2

of supplements." [182]. As Voelker put it in an editorial in the *Journal of the American Medical Association*, the flagship of the medical establishment, in the same year [222]: "The stew of data describing antioxidants' disease prevention potential is coming to a boil. . . . [O]ne thing is certain, vitamin and mineral supplements have entered the ranks of bona fide science." Before we discuss individual diseases in the context of these new developments, we first look at the methods used by scientists to study the role of oxidative stress and antioxidant therapy in those diseases.

how do we study the effects of antioxidants in human disease?

There are two types of epidemiological studies, called the retroactive and the prospective. A third type of study, and often the most instructive, is the double-blind, placebo-controlled method.

In the retroactive epidemiological method, researchers select from the general population a representative sample of, say, ten thousand people. They apply certain criteria to eliminate subjects who have conditions that might complicate the issue (illness, longevity, alcoholism, and so on). They then obtain a dietary history, in which each subject tries to remember what sort of food he or she has eaten over a given number of years. If the focus of study is, for example, the antioxidants vitamins C and E and beta-carotene, researchers then estimate how much of these vitamins the reported diet would contain. They partition the sample into several groups, taking data from those with the highest level of intake (usually the upper fifth) and those with the lowest (usually the lowest fifth). Researchers then ascertain the number of cases of the particular disease under study that have occurred in each group at the two extremes, the highest fifth and the lowest fifth. The results may show that the disease occurs less often in the group with highest vitamin E intake than in that with lowest intake. This would appear to indicate that vitamin E protects against getting that disease.

However, this method has drawbacks. First, it is difficult for most people to remember their diet over many years. People's memories about such matters are notoriously unreliable. Furthermore, experiments have shown that, when people fill out questionnaires, they often unwittingly overestimate their consumption of fruits and vegetables. The second drawback is the "uncontrolled variable" problem. Subjects with a high intake of vitamin E may also have a healthier lifestyle—which consists of engaging in more exercise, eating less fat and drinking less alcohol, being subject to less stress, and so on—than those with a low intake of vitamin E. It may be precisely that combination of a low-fat diet, more exercise, and less stress that accounts for the lower disease rate, rather than the amount of vitamin E in the diet. Although a trial can be designed to control for obvious factors such as alcohol and tobacco use, there is always the possibility of some other factor that researchers had not thought of being the more significant reason for the difference in disease rates between two groups.

Prospective epidemiological studies also operate on a selected population of subjects, but this time researchers follow those subjects for a number of years, measuring their antioxidant intake and sometimes their antioxidant blood levels. At the end of a set period of time, researchers ascertain how many members of the highest and lowest intake groups of the antioxidant under study have developed a particular disease. The advantage of this method is that the researchers get more accurate readings of each individual's antioxidant intake levels and do not have to depend on unreliable memories. The disadvantages are the cost and the long time required to complete the experiment. This method also suffers from the uncontrolled variable problem in that one can never eliminate the effect on the final results of a better lifestyle and other factors not identified when the experiment was designed.

The best way of obtaining information is a double-blind, placebo-controlled study, of which there are two types, preventative and therapeutic.

In a preventative study researchers select a number of apparently healthy subjects using certain exclusion criteria (e.g., below or above a certain age range, the presence of other illnesses, and so on) and divide them into two equal groups, the treatment group and the placebo group, matched for age, sex, and any other factor that might be significant. If all the subjects in the treatment group were older people and all the subjects in the placebo group were younger, then it could never be proved that any apparent effect of the treatment being tested is not simply attributable to the age difference between the groups. Once the matched groups are selected, the treatment group is given the antioxidant to be tested and the placebo group a placebo pill that looks and tastes the same as the antioxidant; the subjects do not know which one they are taking. In fact, no one knows which treatment an individual receives except for one member of the research team. The other members of the team that perform the clinical evaluations do not know the treatment group assignments. Neither the evaluator nor the subject are in the know—hence the term "double-blind." This experimental design ensures that evaluations are not subject to bias. At the end of the trial, it is determined how many subjects in each group develop the disease under study (e.g., heart disease, cancer, or cataract) and how many have died from it. Of course, the trial can investigate more than one disease.

A therapeutic trial entails the same procedures, but the subjects already have the disease in question; the study is directed toward finding out if the treatment improves their condition, or at least slows down the progress of the disease. The advantage of this method is the likelihood that any results are, in fact, attributable to the treatment and not to some other factor. The disadvantage, in the case of antioxidants, is that these agents have a long-term course of action. Thus, to be relevant, a therapeutic trial must last for years rather than for months. Clearly, if a disease condition takes ten years to establish itself, then a clinical trial lasting only one year will not yield any meaningful result. But long-term trials are expensive and a burden on both the investigators and the patients. It is difficult to find subjects willing to wait for years before they can be told what they are taking. Also, in a protracted trial it becomes easier for the double-blind status to be compromised—that is, for someone to discover inadvertently which pill is which.

One thing we know about antioxidants is that they are more likely to help prevent various diseases than to cure them once they have started. In a protracted illness, such as Parkinson's disease, antioxidant therapy is more likely to be effective if it is started before the affected nerve cells in the brain are destroyed. However, such timing is difficult to achieve as these cells begin to die five years before the first clinical symptoms of the disease are manifest. Since there is no existing test to tell which people will develop Parkinson's disease, meaningful therapeutic trials to study the relationship between antioxidants and Parkinson's disease cannot be carried out. A further problem is that many of the studies under consideration in this book test only one, or perhaps two, antioxidants. This is a poor strategy, because antioxidants work together as a synergistic team. A study using large amounts of just one antioxidant may seriously disrupt this synergistic process and lead to unreliable, or even undesirable, results as is detailed later.

A final drawback to therapeutic trials is that, even in a disease known to be caused in part by reactive oxygen species, giving antioxidants by mouth or by injection does not guarantee that they will ever get to the place at which they are needed. The body is not a sack filled with fluid in which medicines simply diffuse everywhere. Rather, the body has highly specific mechanisms, called pumps, that tightly control the uptake of the antioxidant from the gut into the blood, from the blood into the target organ, and once in the organ onto the right place in the right population of cells. For example, vitamin C is soluble in water, but not in fat. Therefore, it could never get inside the fatty cell membrane where fat oxidation takes place. On the other hand, vitamin E is soluble in fat, and so can reach these targets. To circumvent this problem some pharmaceutical companies are busy trying to make fat-soluble derivatives of vitamin C. But even these will not solve every problem. Where the cellular target is in a watery surround, even if surrounded by a lipid (fat) barrier, water-soluble vitamin C will still be needed.

measurement of oxidative stress in the clinic

Because most reactive oxygen species are highly reactive molecules, they do not travel far from where they are made before attacking some biological molecule such as fat, protein, or DNA. Once these reactive oxygen species get attached to the biological molecule, it is no longer possible to detect them without removing the biological molecule from the body and analyzing it for traces of the attached reactive oxygen species. As it is not practical to measure free reactive oxygen species in blood or body tissues in a clinical setting, indirect methods are necessary.

1. Under oxidative stress the metabolism of the cell changes to produce certain stable substances. These can be measured in the breath or in the blood to determine the degree of ongoing oxidative stress. Examples are pentane, MDA (malonyldialdehyde), conjugated dienes, and lipid peroxides. Furthermore, oxidative attack changes the chemical structure of DNA, fat, and proteins. These changes can be detected and measured in the laboratory. Examples are 8-OHαG on DNA, conjugated dienes and MDA/TBA adducts on fats, and carbonyls on proteins.

2. During oxidative stress the small-molecule antioxidants, such as vitamins C and E, when mopping up the excess reactive oxygen species, tend to be used up at a rate faster than the antioxidants can be mobilized and rushed to the scene of action. This means that the blood levels of these antioxidants, especially of vitamins C and E and beta-carotene, will tend to fall as long as the degree of oxidative stress is greater than the capacity of the body to mobilize enough antioxidant defenses to deal with the situation. The blood levels of these antioxidants can be measured, thus indicating how the battle is going.

3. Antioxidant enzymes are synthesized when they are needed. For example, under oxidative stress, levels of SOD, CAT, and GSHpx tend to rise in the blood and body tissues. This is because reactive oxygen species act directly on the nucleus of the cell (via the transcription factor NF-κB mentioned earlier) to start the synthesis of these enzymes. Thus, the local and general antioxidant defenses will be increased. However, in some cases, the disease process may be due to low levels of one or more of these enzymes because something has gone wrong with the cellular machinery making them—in which case blood levels will be low, not elevated.

the role of oxidative
stress in disease
Oxidative stress—damage resulting from an excessive production of reactive oxygen species and/or a failure of the antioxidant defenses—plays an important role in many diseases. Logically, then, antioxidants could potentially play a useful role in therapy. During the initial stage of oxidative stress the body's defenses are mobilized and

levels in the blood and tissues of antioxidant compounds tend to rise. As has been noted, this is in part due to the fact that reactive oxygen species directly stimulate the cell to make more antioxidant enzymes. If the oxidative stress is too prolonged or too severe, the antioxidant defenses tend to be overwhelmed because the small molecules like vitamins C and E simply get used up and their blood levels decrease. Overwhelming oxidative stress may also cause the reactive oxygen species to attack the antioxidant enzymes themselves—which, as proteins, are vulnerable to such an attack—and so diminish their effectiveness. As available antioxidants are consumed without replenishment, with continued oxidative stress and the buildup of reactive oxygen species molecules, cellular damage can occur, leading to the development of disease.

This section describes those diseases thought to be associated with oxidative stress; it includes an account of the results of treatment of these diseases with antioxidants. It should be emphasized that the technical term "oxidative stress" is not the same thing as "stress," as commonly used to denote psychological or life stress. But the term does convey the sense of the continual battle in all living cells between potentially lethal reactive oxygen species and the antioxidant defenses.

The order in which the first twelve diseases is presented corresponds to both the degree of their impact on the population at large and the degree of importance of oxidative stress in their etiology. The rest are given in alphabetical order.

HEART

DISEASE
Coronary artery disease causes heart attacks and is the single major cause of death in the United States. It is caused by the slow buildup of a form of organized blood clot (called an atheromatous plaque) in the walls of the arteries that supply the heart with blood. Many factors combine to cause these plaques, including genes, excess animal fats and cholesterol in the diet, excess low-density lipoprotein (LDL, the bad form of cholesterol) in the blood, lack of exercise, smoking, and obesity. But one major factor related to the others is the oxidation of fats in the blood. As a result of this oxidation the fats in the blood (and elsewhere in the body) become, so to speak, rancid. What implicates LDL

cholesterol in coronary heart disease is the fact that it is more easily oxidized than are normal fats. Oxidized fats stick to the walls of arteries and
are then taken up by certain white blood cells called macrophages (Greek
for "big eater"). Typically, when these macrophages take up a normal (non-
LDL) fat, they become "full" and stop consuming fat. They then move
off to allow another macrophage to take over. In the case of oxidized LDL
fats, this process is altered. When the macrophages take up an oxidized
LDL fat, the switch-off mechanism that signals fullness fails to work; the
cells go on taking up oxidized fat and become overloaded and bloated with
fat to form "foamy cells." These foamy macrophages cease to operate properly and remain in the artery wall to help form the organized clot. When
the clot blocks the normal flow of blood, a heart attack is likely to result.

It is not only the failure of normal macrophage regulation that leads
to coronary heart disease. Increased "stickiness" of the white cells and
platelets in the blood is also involved. The fatty deposits in the wall of
the blood vessels cause white blood cells (including foamy macrophages)
and platelets to stick to them, which starts the process of clot formation.
Anything that increases this stickiness will be a risk factor for heart
attack. However, if these blood cells are not sticky enough, they will fail
to repair small tears in the vessel wall and hemorrhage may result. As with
any regulated process, too much or too little stickiness can lead to health
problems. Vitamin E has been shown to reduce stickiness, especially in
the case of blood platelets, by a mechanism independent of its antioxidant property. It also protects LDL fats against oxidative attack by its
antioxidant property.

A heart attack itself causes profound oxidative stress, with release of
reactive oxygen species especially during the reperfusion stage, when
the blood flow is returning to the damaged heart muscle. Chandrasekar
et al. have shown that, during the reperfusion stage, the levels of messenger
RNAs for proinflammatory cytokines in the heart are raised [32]. This ele-
vation means that a signal was sent to the cell nucleus to manufacture
more cytokines (which are proteins) by means of the DNA-to-RNA system of protein synthesis, of which messenger RNAs are a part. The pur-
pose of this cytokine production is somewhat obscure, as proinflammatory
cytokines only make things worse by increasing the blockage of the coronary artery. At the same time the research shows that a signal was sent

to the nucleus of the heart muscle cell to manufacture more of the key antioxidant enzymes CAT, SOD, and GSHpx (see part 1) so as to protect the heart muscle by counteracting the deleterious effects of reactive oxygen species.

If oxidation of fats is one of the culprits, then antioxidants may be part of the remedy. If antioxidants could slow down the oxidation of the fats that start this pathological process, they might help prevent heart attacks. However, to be effective these antioxidants would have to be fat soluble (see part 1) so as to be able to penetrate to where the damage is—in the fat.

Many experiments in animals and in the test tube have shown that vitamin E will protect cells and fats against fat oxidation induced by reactive oxygen species. Heart muscle is particularly vulnerable to oxidative stress because of its heavy workload and its normal low level of antioxidant defenses [180]. So it is likely to be adversely affected by any interference with its blood supply, as in a heart attack. In the case of any organ that temporarily loses its blood supply, much of the damage is done by reactive oxygen species during the reperfusion period, when the blood is entering the organ again.

CLINICAL DATA AND MAJOR RESEARCH STUDIES. A look at the clinical data in humans and a review of the major research studies will help us to understand whether antioxidants can prevent or alleviate coronary heart disease, which I henceforth term "heart disease" with the understanding that this includes only coronary heart disease and not conditions like heart failure or irregularities of the heartbeat. The data from the thirty-four most prominent studies on the relationship between antioxidants and heart disease is presented. These studies are numbered for ease of reference; the reader can follow the analysis of how these studies relate to each other. You will note that there are disagreements in these reports between the authorities who carried out the researches, and you may wonder why. After all, it did not take long to discover that penicillin is effective. However, in the case of these chronic and complicated diseases, arriving at firm conclusions is more difficult. Reliable judgments must be based on all the research work that has been carried out and on the recommendations made by all the experts in the field.

In the first eight studies the investigators examined the relationship between the level of ingestion of antioxidant vitamins and the incidence of heart disease.

(1)   In the Nurse's Health Study, Stampfer and Rimm measured the intake of vitamins C and E and beta-carotene in 87,245 American female nurses over an average period of eight years—a truly formidable undertaking [207]. They divided the group into the fifth part (quintile) with the highest intake and the fifth part with the lowest intake. The average level of vitamin E taken by the highest quintile was 208 mg per day and the average level taken by the lowest quintile was 2.8 mg per day. They found that the higher level of intake of vitamin E was associated with a 34 percent reduction in heart disease. A higher consumption of beta-carotene was associated with a 22 percent decrease. If the intake of both these antioxidants was high there was a 50 percent reduction. Vitamin C supplements had no effect. They also noted that vitamin E was effective only if it was given as a vitamin E supplement and not as merely one ingredient in a multivitamin pill. They explained this on the basis of the fact that vitamin E supplements typically contain more than 100 mg per day whereas the multivitamin pills contain only 30 mg per day or less. This is evidence that the current recommended daily allowance (RDA) for vitamin E may be too small to play a role in preventing heart disease.

(2)   The same team then looked at a population of 39,910 men over a four-year follow-up period using the same methods [176]. The median intake of vitamin E in the case of the highest quintile was 419 mg per day and in the case of the lowest quintile it was 6.4 mg per day. Again they found that the higher level of vitamin E intake was associated with a significant reduction in the rate of development of heart disease; vitamin C had no effect. In this study beta-carotene was also effective but only in smokers. This finding with beta-carotene might have been due to the shorter term of this study as compared with their first study. This suggests that the therapeutic effects of beta-carotene may take a long period of time to develop. There may also be a gender difference involved, since this study involved only men whereas the previous study involved only women. The authors point out

that the "healthier lifestyle" explanation for these results was unlikely to be right, as the subjects who took high levels of vitamin C supplements had equally healthy lifestyles as those who took high levels of vitamin E supplements, but obtained no benefit. In a later paper the same authors state that at least 100 mg per day of vitamin E intake is required to produce an effect in lowering the rate of heart disease [207].

(3) The third study was carried out on a population of 34,486 post-menopausal women who were free from heart disease at the start of the trial [114]. The investigators measured the intake of antioxidant vitamins and found that, if vitamin E was obtained from food (mainly from nuts, seeds, margarine, and mayonnaise), there was a strong protective effect against heart disease. There was no protective effect if the vitamin was taken as a multivitamin supplement. The authors explained this result also as owing to the fact that the dose of vitamin E in the multivitamin supplements used by these subjects was too low. Again, vitamin C had no effect.

(4) The Physician's Health Study is another major study, this time confined to beta-carotene. For twelve years, 27,071 American male physicians each received a small dose of beta-carotene (equivalent to two carrots per day) [82]. The investigators reported no benefit for preventing heart disease from this level of beta-carotene intake.

(5) A similar study carried out in Basel, Switzerland, found that eight years of beta-carotene intake showed no clinical benefit; but if the length of the study was increased to twelve years, there was a significant reduction in the incidence of heart disease [66]. The researchers stress heart disease's slow rate of development and, if one is to get a true estimate on the effectiveness of any procedure, the need for long trials rather than short ones.

(6) In the Scottish Heart Health Study it was found that a high level of intake of vitamins E and C and beta-carotene appeared to protect men, but not women, against heart disease [216]. This study used a sophisticated statistical method (multivariant discriminant analysis) for analyzing the results; the actual levels of vitamin E intake involved are not mentioned.

(7) The Lipid Research Clinics–Coronary Primary Prevention Trial studied 1,899 men aged 40 to 59 and measured total carotenoids (not just beta-carotene, which contributes only 25 percent of the total carotenoid intake of the diet) [146]. The results showed a strong protection against heart disease for men who had never smoked. The authors also stress that the duration of such studies must be at least ten years for valid conclusions to be drawn.

(8) One of the longest longitudinal trials ever conducted studied 1,556 men over a period of twenty-four years [158]. The results showed that high levels of intake of beta-carotene and vitamin C led to a 20–30 percent decrease in the death rate from both heart disease and cancer.

In studies 9 to 13, the researchers investigated the relationship between blood levels of antioxidant vitamins and the incidence of heart disease.

(9) Gey has reported the results of the MONICA project carried out by the World Health Organization [65]. This project studied the relationship between blood levels of vitamin E and death rates from heart disease in sixteen European countries and found that vitamin E had a strongly protective effect. When given in sufficient amounts, vitamin E worked better in combination with several antioxidant "helpers" than on its own. As we have seen, antioxidants (particularly vitamins B_3 or nicotinamide, E, and C) work together as a team and need to be given as a team, in sufficient quantities, to be most effective. Most multivitamin tablet formulations contain insufficient amounts of vitamin E.

(10) Riemersma et al. measured blood levels of vitamins E and C in apparently healthy middle-aged men in four locations—two in Finland, one in Scotland, and one in southern Italy [174]. In the first three locations there is a very high incidence of heart disease; in southern Italy heart disease is less prevalent. Blood levels of vitamins C and E were low in Scotland and in one of the Finland locations and high in Italy. Italians typically eat any more fresh fruits and vegetables than do Scots and Finns. But in Karelia, the other Finland location, vitamin C levels were not low, yet the rate of heart disease was

high. The authors suggest that this apparent discrepancy might be due to the fact that the Karelians were in general more obese and had higher blood pressure than did the other three groups—both factors that predispose to heart disease. Indeed, dietary studies show that Finns have the highest intake of fat in Europe. This again illustrates the important point that other factors (such as blood pressure and obesity) may complicate the interpretation of the effects of antioxidants.

(11) Another study compared two groups of young people, one from Naples, Italy, and the other from Bristol, England [161]. The Italians had higher blood levels of vitamin E and much higher levels of beta-carotene. Although the groups ate the same quantity of vegetables, the Italians consumed considerably more tomatoes and olive oil. The Italians had much lower rates of fat oxidation (as measured by plasma levels of conjugated dienes and lipid peroxides, which are products of lipid oxidation). Tomatoes, which contain the important antioxidant lycopene, and olive oil, which also contains many antioxidants, are key ingredients of the Mediterranean diet, which many studies have shown offers powerful protection against heart disease and some forms of cancer.

(12) Singh et al. studied impoverished industrial workers who ate a poor diet and had high exposure to toxic fumes derived from diesel engines and heavy metals such as copper and lead [196]. Those subjects who developed heart disease had lower blood levels of vitamins E and C and of beta-carotene than did those who did not develop heart disease.

(13) Middle-aged men in Lithuania have four times as many heart attacks as middle-aged men in Sweden. To find out why, Kristensen et al. studied one hundred people, fifty from each country [111]. There were no differences between the two groups in the ordinary risk factors such as blood cholesterol, smoking, obesity, and high blood pressure. But there were two significant differences: the Lithuanians had low-density lipoproteins (bad cholesterol) that were more easily oxidized, and they had lower blood levels of some antioxidants, such as beta-carotene, lycopene, and, interestingly, gamma-tocopherol (a relative of vitamin E). Levels of alpha-tocopherol (vitamin E) were the same. The investigators suggested that their results might be due to the fact

that Lithuanians ingested more polyunsaturated fats but concluded that their antioxidant status was also important. The authors quote another Swedish study that showed that men with coronary heart disease had low blood levels of gamma-tocopherol but normal levels of alpha-tocopherol. This underscores the importance of gamma-tocopherol in the diet and suggests that, in future, such studies should pay as much attention to gamma-tocopherol as to its more famous relative alpha-tocopherol.

Studies 14 to 16 concentrated on vitamin C alone.

(14) The First National Health and Nutrition Examination Survey (NHANES 1) cohort study was based on a ten-year follow-up on the diet of 11,348 American adults aged 25 to 74 [52]. The results indicated that a high dietary intake of vitamin C appeared strongly to protect men against death from heart disease (as well as death from cancer and death from all causes); but women were only weakly protected. However, this study has been criticized on technical grounds by Herbert [83]. Enstrom has replied to this criticism [51].

(15) Toohey et al. studied 172 African American Seventh-Day Adventists (whose religious practice prohibits smoking tobacco and drinking alcohol) [217]. The results showed that a high intake of vitamin C was strongly associated with a lower rate of fat oxidation (probably because of the way vitamin C "helps" vitamin E, as described in part 1). The researchers concluded that their results were probably not due merely to eating more fruits and vegetables, mainly because they felt that they had not eliminated the possibility of the influence of other factors, such as a better lifestyle.

(16) Nyyssönen et al. tested 1,605 men aged 42 to 60 in a long-term study in Finland of the effects of vitamin C deficiency [149]. The apparently healthy subjects entered the study between 1984 and 1989 and were followed up until 1996. Some ninety subjects had a blood level lower than the cutoff point that indicates a deficiency state of 2 mg/L. These subjects showed a highly significant increase in the number of heart attacks when compared with the rest of the subjects who had normal blood levels of vitamin C. This was so even allowing for other factors such as the amount of saturated fats, carotene,

fiber, and tea consumed. However, further increases in the blood level of vitamin C above the limit of 2 mg/L conferred no further protection. This suggests that vitamin C supplements by themselves would not have been of help. However, it is a matter of concern that so many Finns were deficient in vitamin C. Physicians should bear in mind the possibility of a previously undetected vitamin C deficiency when evaluating a subject's risk of having a heart attack.

The protective effect of antioxidants was estimated in studies 11 and 15 by measuring the levels of fat oxidation that predisposes toward heart attacks; in the other studies it was estimated by monitoring the development of heart disease itself. We can sum up the foregoing review of the major epidemiological studies as follows.

INTAKE STUDIES
- High dietary levels of vitamin E alone protected against the development of heart disease in three studies (1, 2, and 3). There were no studies with negative results.
- Vitamin C was protective or partly protective in two studies (14 and 15) but had no effect in three studies (1, 2, and 3). Its main effectiveness may be in people who have an actual vitamin C deficiency state.
- Beta-carotene was protective in study 5 but not in study 4.
- The combination of vitamins E and C and beta-carotene was protective in study 6.
- Total carotenoids were protective in study 7.
- The combination of beta-carotene and vitamin C was protective in study 8.
- Vitamin C was protective in vitamin C deficient individuals in study 16.
- In two studies (6, 14) men were reported to have a more favorable clinical course than women.

BLOOD LEVEL STUDIES
- High blood level of vitamin E alone was protective in study 9.
- High blood levels of vitamins E and C were protective in study 10.
- High blood levels of vitamin E and beta-carotene were protective in study 11.

- Low levels of vitamin E, vitamin C, and beta-carotene were a risk factor in study 12.
- Low levels of beta-carotene, lycopene, and gamma-tocopherol constituted a risk factor in study 13.

Collectively, this epidemiological data strongly supports a protective role for vitamin E in the prevention of coronary heart disease, if given at an adequate dose (over 100 mg per day). It is better to administer vitamin E together with its supportive antioxidants. The evidence for beta-carotene's role is weaker, but total carotenes appear to be more effective than beta-carotene by itself. The evidence for any protective role of vitamin C administered by itself is unconvincing, except in people who have an actual vitamin C deficiency—possibly because heart disease is the result of fat oxidation and vitamin C, a water-soluble antioxidant, does not penetrate a fatty environment. However, if vitamin E is indeed protective, then vitamin C may be needed in its role as a helper of vitamin E. Thus we can expect vitamin C to be useful when given together with vitamin E, but not when given by itself.

FIVE DOUBLE-BLIND, PLACEBO-CONTROLLED STUDIES

(17) Stephens et al. have recently reported one of the first results of such a trial—the CHAOS study carried out in Cambridge, England [201]. There were 2,002 subjects who were studied for an average of seventeen months. Half were given vitamin E (800 mg per day), and the other half received a placebo capsule. In the test group blood levels of vitamin E rose substantially, showing that it was being absorbed adequately. The results showed a significant fall in the number of heart attacks in the group that had been given the vitamin E. The number of deaths did not decrease, but the investigators noted that most of these occurred at the beginning of the trial, so the vitamin E might not have had time to work. Other studies have demonstrated that benefits from administration of vitamin E need time to develop.

(18) Another double-blind study showed that vitamin E (400 mg per day) was protective against heart attacks [208]. The protection was increased in the case of those subjects who also took aspirin (325 mg) daily.

(19) The Alpha-tocopherol–beta-carotene (ATBC) study, which was primarily an investigation of cancer [5], also reported on the incidence of heart attacks among the subjects [170]. The study involved 1,862 male smokers in southern Finland; each had had one heart attack. The subjects were divided into four groups. Group A received 50 mg per day of vitamin E; group B received 20 mg per day of beta-carotene; group C received both; and group D received neither. After five years the subjects in group A, who had taken only vitamin E, had no difference in fatal heart attacks as compared with the control group D, but showed a 38 percent reduction in nonfatal heart attacks. But both groups who were given beta-carotene showed significant increases in fatal heart attacks. In an editorial in the issue of the *Lancet* [209] that reported the results, Stephens, a member of the CHAOS team, pointed out that the ATBC study used only one-tenth the dose of vitamin E that the CHAOS study had used; moreover, the ATBC investigators had used synthetic vitamin E whereas the CHAOS group had used natural vitamin E, which is chemically different (as we saw in part 1). Stephens concluded that the bulk of the evidence supports the use of vitamin E, but not beta-carotene, to treat coronary heart disease. Furthermore, in study 10 reported above, the southern Finns had an aberrant result, as they had high blood levels of vitamin C but also a high level of heart attacks; in other studies high blood levels of vitamin C are associated with a low incidence of heart attacks. The confounding factor in that study might have been the subjects' very high consumption of fat. Interestingly enough, another study of Lapps living in northern Finland showed that they had very low levels of heart disease [129]. Lapps have a quite different diet from southern Finns'. The researchers attributed the Lapps' low levels of heart disease to their diet, which provides rich sources of vitamin E, albumin, and selenium.

(20) The Established Population for Epidemiological Studies in the Elderly examined the role of antioxidant supplementation (vitamins E and C) in a population of 11,178 people aged 67 to 105. Supplements of vitamin E alone were significantly correlated with a lower mortality rate from heart attacks [128].

A summary of double-blind, placebo-controlled studies is as follows:

- Vitamin E appeared to reduce the mortality rate from heart attacks in one study (20) but not in two others (17, 19).
- Vitamin E appeared to reduce the number of heart attacks in all four studies (17–20).
- Vitamin C was not tested by itself.
- Beta-carotene appeared to increase the mortality from heart attacks in one study (19).

In studies 21–26 the subjects were patients actually undergoing an acute heart attack.

(21) Singh et al. studied 109 cases of acute myocardial infarction and 182 controls [197]. They found that the plasma levels of vitamins A, C, and E and of beta-carotene fell during the attack. Plasma levels of oxidized fats rose.

(22) Patients following a heart attack had a significantly higher death rate if their blood total-antioxidant capacity (excluding albumin and uric acid) was low [144].

(23) In a randomized, double-blind controlled study of 125 patients with a myocardial infarction, 63 of whom were given antioxidants (vitamin A 50,000 international units; vitamin C 1 G; vitamin E 400 mg; and beta-carotene 25 mg) for twenty-eight days and 62 of whom were given placebo, the patients who received the antioxidants had less severe heart attacks (smaller infarct size, less pain, improved heart function, and lower levels of oxidized fats in the blood) than did those who had received placebo [198].

(24) & (25) Chamiec et al. measured the effects of vitamin E (600 mg per day) and vitamin C (600 mg per day) on the electrocardiogram (EKG) after an acute heart attack in a placebo-controlled study [31]. They found that the placebo group had a number of signs in the EKG typical of a heart attack. These EKG changes were not seen in the vitamin-treated group. In a similar study using another antioxidant (N-acetyl cysteine or NAC), the patients given the antioxidants showed less impairment of heart function and less fat oxidation than did the patients given the placebo.

THE EFFECT OF ANTIOXIDANTS ON PHYSICAL CHANGES IN ARTERIES

(26) Coronary angiography is an X-ray method of looking at the arteries of the heart. Hodis et al. used this method on 156 men given supplements of vitamins C and E after coronary bypass surgery [88]. The results were that in mild and moderate cases the vitamins led to a significant slowing down of the disease process in the arteries as shown by the X-ray pictures. However, this was not true in severe cases. The slowing down of disease was largely due to the intake of over 100 mg per day of vitamin E.

(27) A similar investigation was carried out on the carotid arteries in 1,187 people [21]. The investigators used ultrasound to measure in the living person the degree to which atherosclerosis had blocked the artery. High blood levels of vitamin E were associated with less artery blocking. Also, there were more signs of fat oxidation in the blood in those people with low blood levels of vitamin E, selenium, and beta-carotene. The investigators concluded that their results support a protective role of vitamin E in heart disease.

Studies 24–27 yield important findings that provide hard objective evidence of the benefits of antioxidant treatment in heart disease. Furthermore, the studies in this section (17–27) without exception demonstrate that vitamin E at a dosage of around 400 mg per day helps to prevent heart attacks and lessens the effects of an acute heart attack. Again, vitamin C seemed to have little effect except as a helper of vitamin E.

(28) Patients on renal dialysis have five times the normal rate of heart attacks, possibly because the dialysis procedure removes the key antioxidants vitamin C and uric acid from the blood. Such patients urgently need to increase their antioxidant intake.

Animal studies have shown that vitamin Q_{10} protects the lining of the blood vessels in the heart against oxidative attack [227]. Several studies have focused on the antioxidant vitamin Q_{10} in humans, of which two are reviewed here.

(29) Kuklinski et al. studied 61 cases of acute myocardial infarction (heart attack) for one year [113]. Half were given vitamin Q_{10} plus 500 µg selenium, and the other half placebo. In the experimental group

none showed any abnormality of the EKG, with no deaths from heart disease. In the control group given placebo 40 percent showed an abnormal EKG, and there were six deaths from myocardial infarction. The authors advise antioxidant treatment for all cases at risk.

(30) A similar but uncontrolled study showed that vitamin Q_{10} (75–600 mg per day) in 424 cardiac patients (average follow-up time, 17.8 months) was correlated with clinical and electrocardiographical improvement [116]. These authors cite several previous double-blind, placebo-controlled studies which show vitamin Q_{10} to be "safe and effective."

THE FRENCH PARADOX. Nutritional experts have recognized for years what they call the "French paradox." Although the French have a diet rich in fats, cholesterol, and all sorts of delicious foods that are supposed to be very bad for you, they also have a low rate of heart disease. Some nutritionists have suggested that this is because of the red wine that they drink [69, 70]. (Of course, they also eat a diet rich in protective fruits, vegetables, herbs, garlic, and olive oil.) Red wine contains powerful antioxidant flavonoids (in particular quercetin, rutin, reservatol, and catechin). White wine contains lesser amounts. These chemicals are also found in tea, onions, and apples. However, it is possible that the French paradox may also be due in part to the higher intake of vitamin E, especially in the form of sunflower seed oil.

(31) Hertog et al. in Zutphen, Holland, studied 805 men aged 65 to 84 for five years [86]. They measured the total flavonoid intake, 65 percent of which came from tea, 13 percent from onions, and 10 percent from apples. They found that a high flavonoid intake was associated with a lower death rate from all causes. Three other studies reviewed by Hollman et al. (the Netherlands Cohort Study, which involved 120,850 men and women, and two studies in Finland, one of which involved 5,130 men and the other women and 550 men) all showed a protective effect of high dietary flavonoid intake against heart disease but not against cancer [91].

It has been suggested that if everyone drank two glasses of red wine a day the rate of heart disease would fall by 40 percent. This may be somewhat of an exaggeration, but certainly flavonoids are important in the diet. All four studies on flavonoids here described found them to be protective against heart attacks.

(32) Another recent study, which involved researchers in ten European countries, has shifted the focus of interest onto lycopene (the main antioxidant in tomatoes) [110]. Instead of the usual blood levels these researchers measured the level of various fat-soluble anti-oxidants in the body fat (adipose tissue). This is a logical place to look for compounds that are soluble in fat rather than in watery blood. They found, first, that lycopene and beta-carotene levels in blood are about the same, but that fat contains much more lycopene. Next, they found that high fat levels of lycopene were associated with a lesser risk of heart attacks. This was not so for alpha- or beta-carotene. They also comment that lycopene is a much better antioxidant than beta-carotene. It is possible, therefore, that tomatoes are more protective than carrots.

(33) Nitroglycerine is a drug that is commonly used in the treatment of angina. One of its drawbacks is that tolerance develops to its action— that is, doses that are initially effective do not remain so. One ran-domized, double-blind, placebo-controlled trial has found that vitamin E supplements prevent the development of this tolerance [227]. The researchers therefore recommend the use of vitamin E supplements in patients who take nitroglycerine.

(34) A study comparing smokers and nonsmokers reports that people who smoke cigarettes have low levels of vitamins C and E but normal antioxidant enzymes in their red blood cells. When smokers have a heart attack, their bodies respond with more severe changes in anti-oxidant status (lowered levels of antioxidant enzymes SOD, GSHpx, and CAT, and antioxidant vitamins A, C, E, and GSH) than those of nonsmokers. The authors of the study recommend that smokers take antioxidant supplements, in particular vitamins C and E.

BLOOD PRESSURE. Finally, there has been one randomized, double-blind, cross-over study of the effect of antioxidants on blood pressure. Some 30 percent of adults in the West suffer from elevated blood pressure.

Galley et al. in Scotland selected thirty-eight patients with high blood pressure and seventeen patients from the same clinic with normal blood pressure [61]. Each patient received an antioxidant cocktail (or placebo) for eight weeks, followed by a two-week washout period, and then placebo (or antioxidant cocktail) for another eight weeks. The cocktail consisted of vitamin C (500 mgs), vitamin E (600 mgs), beta-carotene (30 mgs), and zinc (200 mgs). Blood pressure was significantly lowered in both groups of patients, but more so in those with high blood pressure. The latter group also showed an increase in urinary excretion of nitrite, a metabolite of nitric oxide. The researchers hypothesized that the blood pressure was lowered because the antioxidants protected nitric oxide (which lowers blood pressure by dilating blood vessels) from oxidation by the superoxide ion. Since high blood pressure predisposes to heart disease and strokes, the researchers suggested that this vitamin cocktail might be effective in reducing the incidence of these diseases. A study of the effects of vitamin Q_{10} on hypertension showed that it had a positive clinical effect [117]. Blood pressure level was reduced and there was less need for antihypertensive drugs.

These studies provide strong evidence that vitamin E should be taken at a dose of at least 400 mg per day by anyone wishing to decrease the likelihood of having a heart attack. This dose cannot easily be obtained from diet alone, as one would end up eating too much fat. The evidence indicates that this dose of vitamin E is safe, except in the case of people with certain blood coagulation disorders; it may result in a small increase in the risk of hemorrhagic stroke, as detailed in part 3. Along with the vitamin E one should also take an adequate amount of its helper antioxidants. However, beta-carotene by itself should be avoided, especially by people who have had one heart attack. Why is the efficacy of this dose of vitamin E not widely accepted by the authorities, including many physicians and other health professionals? Given the evidence, it is rather surprising that many experts still maintain that the official RDA of vitamin E of 20 mg per day—twenty times smaller than the amount needed to confer a protective effect against heart disease—is sufficient to cover all our health needs. The motivations and implications of the current policies will be considered later.

CANCER   Cancers can originate in almost all the cell types in the body, and the development of cancer is an enormously complicated process in which many biochemical systems play a role. Prominent among these systems are those that damage the cell's DNA. Many agents can damage DNA: ultraviolet light, radiation, cancer-producing chemicals such as tar, and reactive oxygen species molecules. A large number of animal experiments show that reactive oxygen species can induce cancers in various systems and that antioxidants can be protective. Almost all cancer cells have low levels of the antioxidant enzymes CAT and GSHpx and have an abnormal regulation of these enzymes [150]. As a consequence, they are more vulnerable to oxidative stress and further DNA damage than are normal cells. Indeed, the cancer cells themselves show much chemical evidence of oxidative stress. With defective antioxidant enzymes, these cells cannot properly handle hydrogen peroxide (H_2O_2), which process, according to researchers, "offers tremendous potential for cancer therapy" [150]. It is possible that new drugs may be developed that can detoxify the hydrogen peroxide. Moreover, it is possible, but not yet established, that antioxidants may affect biochemical mechanisms other than damage to DNA that are involved in the formation and growth of cancers. These may include the complex mechanisms that govern the differentiation and growth of cells.

What is now required are experiments in humans that show whether or not antioxidants, either in a healthy diet rich in fruits and vegetables or given as supplements, actually help to prevent or alleviate cancer in people.

The Work Study Group on Diet, Nutrition, and Cancer of the American Cancer Society has estimated that about one-third of the half-million deaths each year from cancer in the United States are a consequence of diet [236]. Thus, more than 150,000 deaths a year could be prevented by a simple change of diet, to say nothing of the resulting enormous reduction in health costs. The Work Study Group recommends the following steps to lower cancer rates: eat less but consume a more varied diet; eat more fruits, vegetables, whole grain cereals, legumes, and nuts; lower fat intake; drink less alcohol; avoid smoked, salt-cured, and nitrite-preserved foods; and take antioxidant supplements (especially vitamin E) "in certain cases."

There has been an enormous volume of clinical research into the role of diet in cancer. Block et al. at the National Cancer Institute reviewed 156 epidemiological studies of the relationship between a diet rich in fruits and vegetables and various forms of cancer [17]. They reported that for most cancers such a diet cuts the cancer rate in half. Especially favorable sites are the lung (after control for smoking), in which 24 out of 25 studies had positive results; esophagus, mouth, and larynx (24 out of 25 positive); pancreas and stomach (26 out of 30 positive); colon, rectum, and bladder (23 out of 28 positive); and cervix (11 out of 13 positive). Clearly, fruits and vegetables contain some agents that protect against cancer. Block et al. concluded that the healthy lifestyle hypothesis was unlikely to account for the data. They also stressed the great importance of the helping action of the large number of different antioxidants in the diet.

Tavani and La Vecchia of the Mario Negri Institute in Milan, Italy, have reviewed a large number of studies of the effect of diet in a Mediterranean population and concluded that fruits and vegetables offer a strong protective effect (61–87 percent) against many forms of cancer (better protection in the higher respiratory and digestive tracts than in the lower tracts), but none against cancers of the lymphatic system [213]. The evidence suggested that raw vegetables were better than cooked, and vegetables were better than fruit.

Flagg et al. have summarized what they considered to be all the adequately controlled and conducted studies published between 1985 and 1992 on the relationship between dietary intake and blood levels of the antioxidants beta-carotene and vitamins C and E, and various forms of cancer [55]. They estimated the strength of the protection and whether there was a consistent relationship between higher blood levels of the antioxidants and increased protection (called a "dose/response" relationship). The results indicated that the antioxidants provided good protection against cancer of the lung and upper respiratory and digestive tracts, some protection against cancer of the colon and cervix, but none against breast and prostate cancer. It is generally thought that breast and prostate cancers are mainly caused by a disturbance in hormonal control; thus, the lack of protection afforded by antioxidants is not surprising. However, one study of 47,894 subjects in Boston, which measured carotenoids in the diet, found that a diet high specifically in lycopene (from tomatoes or tomato paste) seemed to protect against the development of prostate cancer during the four years of the study [67].

Diplock has also reviewed a large number of studies on the relation of antioxidant dietary factors and cancer [43]. He found that antioxidants were protective in cancers of the lung, upper respiratory and digestive tracts, stomach, bladder, pancreas, cervix, and ovary, but not in cancers of the breast and prostate. He also concluded that for colorectal cancer some ingredient of the diet seems to be important, but it might be fiber rather than antioxidants.

Some reports of prospective studies at first sight would appear to paint a less optimistic picture. A 1996 issue of the *New England Journal of Medicine* carried two important papers and an editorial that have had widespread repercussions. The first paper related to the CARET study (beta-carotene and Retinol Efficacy Trial) [154]. This was a four-year double-blind, placebo-controlled study of 18,314 males and females with a high cancer risk from smoking or exposure to asbestos. The dosage administered to subjects was 30 mg of beta-carotene (equivalent to four carrots a day) plus 25,000 international units of retinol (vitamin A). The results were a 28 percent higher incidence in the treatment group than in the placebo group of lung (but not other) cancer, 17 percent higher incidence of mortality from all causes, and a 28 percent higher incidence of mortality from heart disease—leading to early termination of the study. However, the authors admit that they could not tell whether the beta-carotene or the vitamin A (or the combination) was the culprit. They concluded that it would be unwise to give large doses of only one dietary antioxidant as this might lead to a serious imbalance of the important helping action of the other antioxidants. This is quite plausible because of the important synergistic action of the antioxidant vitamins. The authors also criticized many of the epidemiological studies reporting an apparent protective effect of fruits and vegetables in the diet because they ignore, or fail to control properly for, other variables, such as the level of intake of fats and red meat, exercise levels, lifestyle, and so on, that might actually have been responsible for the reported results. Many studies have, in fact, done their best to control for such variables.

The second paper reported on the current status of the Physician's Health Study [82], which has lasted for twelve years and has involved 27,071 American male physicians (11 percent smokers; 35 percent former smokers; 54 percent nonsmokers) given a supplement of beta-carotene equivalent to two carrots a day. The results showed no evidence of any benefit for cancer, heart disease, or death rate, but, unlike the results of the CARET study, there was no increase of lung cancer even among the smokers. The conclusion was that beta-carotene (at this low dose) by itself is useless, but that evidence from other studies indicates that vitamin E remains promising. The investigators also state that the increased lung cancer risk reported by the CARET and ATBC (see below) studies is not consistent with the dietary evidence, and that the cancer-promoting action (if real) of beta-carotene may be confined to heavy smokers.

Beta-carotene induces blood vessels to grow; because cancers need new blood vessels in order to expand, this could explain why in some cases beta-carotene might promote tumor growth [191]. Beta-carotene, it is recommended, should not be given to people who have a high carcinogen intake, such as smokers. Another hypothesis has been developed to explain why beta-carotene may lead to the increase in lung and prostate cancer reported by the studies reviewed above [112]. The active form of vitamin D has an anticancer effect in certain tissues, which include lung, prostate, and colon. Beta-carotene can interfere with vitamin D synthesis. Furthermore, beta-carotene is stored in the skin, and this may block the stimulating effect of ultraviolet radiation on vitamin D synthesis in the skin. In the ATBC trial in Finland, 345 of the subjects developed yellowing of the skin. This supports the hypothesis that giving excess levels of one antioxidant may induce a shortage in another antioxidant, thus underlining once again the need to give such antioxidants in a balanced formula. So, in smokers, it might be wise to add vitamin D to beta-carotene supplements. It has also been suggested that the reason why beta-carotene does not work in smokers is that these people have a low vitamin C status and, as we saw, vitamin C is involved in recycling vitamin E [20]. Thus, on this hypothesis, smokers need vitamins C and E rather than beta-carotene.

A third negative paper by the ATBC (Alpha-Tocopherol, Beta-Carotene Cancer Prevention Study Group) in Finland, in which 29,137 males who were heavy smokers were given 20 mg of beta-carotene and 50 mg of vitamin E per day for an average of six years in a double-blind, placebo-controlled study, reported an 18 percent increase in the incidence of lung cancer, more prostate cancer, and an 8 percent increase in deaths from all causes in the vitamin group [5]. However, a later report of this study qualified these results to some degree by admitting that their findings conflict with the dietary evidence [13]. Additional findings reported in this study showed that those subjects with higher serum levels and dietary intake of beta-carotene and vitamin E at the start of the study had a lower risk of developing lung cancer. Furthermore, the vitamin E supplements appeared to protect against prostate cancer (down 34 percent) and colorectal cancer (down 16 percent). The people who got lung cancer tended to be those who also drank more heavily. The researchers concluded that longer, even lifetime, supplementation will be needed to combat such a slowly developing process as cancer.

The editorial that accompanied these reports in the *New England Journal of Medicine* argues (i) that beta-carotene as a sole supplement for cancer prevention is contraindicated, (ii) that the value of antioxidant supplements in general remains unclear, and (iii) that the protective effect of diet (especially the strong evidence for fruits, vegetables, and grains) remains important. But the editorial concluded that it is an open question as to whether the antioxidants in these foods are responsible. The conclusion drawn from this apparent debacle by the Director of the National Cancer Institute, which funded these studies, was: "We do not know how to replace a healthful diet and a healthful life style with simple pills" [74].

Both of these editorial comments seem to me to be overreactions. When all the evidence is considered, rather than isolated sections of it, several points about antioxidants become clear. As we have already had cause to note several times, antioxidants should not be given singly because of the strong evidence of their team action. Because overloading with one may upset this delicate balance and lead to bad results, one should be skeptical about single antioxidants for sale on store shelves. The evidence clearly shows that antioxidants should not be consumed singly but only as well-balanced mixtures of many antioxidants. But it is important to ensure that the antioxidant mixture contains enough of the key antioxidants, in particular vitamins E and C. Many mixtures on sale contain insufficient quantities of these antioxidants. However, if a doctor measures antioxidants in a patient's blood and finds that one or two are low, it is all right to take supplements of these under the doctor's direction (more on this later). In fairness it must be pointed out that these facts were not known a decade or so ago when the National Cancer Institute trials were designed. No one today should design a prospective trial based on just one or even two antioxidants.

It would be wise not to give beta-carotene by itself to heavy smokers. It is also unreasonable to expect that antioxidants could reverse the bad effects of many years of heavy smoking (an average of thirty-nine years in the Finnish study). Moreover, the subjects in the Finnish study were given only 50 mg per day of vitamin E (in addition to the beta-carotene), whereas the effective dose would seem to be ten times as much. In addition, they were given synthetic vitamin E, which is chemically different from natural vitamin E, as we saw in part 1. It must also be remembered that both heart disease and cancer develop slowly and progressively over a long period. The value of antioxidant treatment could better be tested by giving it early on in the pathology and maintaining the effective dosage for a longer period. Omenn has also drawn attention to these flaws in all of the major ATBC, PS, and CARET studies: choice of high risk-subjects, choice of a single antioxidant, too low a dose, and too short a trial [153]. In a textbook on antioxidants Packer and Diplock state that this study should be "viewed with caution" [155]. As the Carotenoid Research Interactive Group concluded, "Although carotenoids should not be considered universal preventers of disease, it is equally inappropriate to downplay beta-carotene as a 'passing fad' " [28].

In 1996 beta-carotene had a piece of good news for a change. Omage et al. gave a group of normal women a diet very low in beta-carotene for sixty-eight days [152]. They found that the measures for fat oxidation in the blood (conjugated dienes) were raised significantly at the end of the trial. They concluded that this result "supports the supposition that beta-carotene may have an important role in antioxidant defense."

Four studies published in the 1990s measured blood levels of antioxidants with respect to cancer risk.

(1) Knekt et al. in Helsinki studied 36,265 Finns with an eight-year follow-up [107]. They found that low plasma levels of vitamin E were associated with a 1.5 increased risk for cancer, particularly gastric cancer. In other words, high serum levels of vitamin E appeared to protect against cancer.

(2) Stähelin et al., in a Swiss study of 2,974 males with a twelve-year follow-up, showed that low levels of beta-carotene correlated with higher risks for lung and stomach cancer [205]. In this study beta-carotene appeared to protect against stomach cancer.

(3) In a study carried out in Maryland, Comstock et al. took serum samples from 25,802 people from 1974 to 1975 [35]. These samples were kept frozen until 1989, by which time 436 of these individuals had developed cancer. Their serum antioxidant levels were compared with those of 765 matched controls from the group who had not developed cancer. The results were that high beta-carotene and vitamin E levels were associated with a strong protective effect against lung cancer. Lycopene (the antioxidant from tomatoes) showed a strong protective effect against pancreatic cancer. However, this study was based on only one blood-level measurement of nutrients, and these blood levels are liable to fluctuate according to the diet taken over the period before the measurement. In another study plasma levels of antioxidants (vitamins C and E, urate, and glutathione) were shown to be low in gastric carcinoma patients, and the measures of fat oxidation were higher. But we cannot deduce from these results whether the low plasma levels of antioxidants contributed to the cause of the cancer or were a result of it.

(4) Last, Ziegler et al. measured the levels of six carotenes in a population of Japanese Hawaiians in relation to the risk of developing cancer of the lung, esophagus, larynx, throat, and mouth [244]. They found that low levels of alpha-carotene, beta-carotene, beta-cryptoxanthin, and lycopene correlated with a high incidence of lung cancer. This was particularly true in the case of alpha-carotene (more than beta-carotene). Other cancers correlated with low levels of alpha-carotene, beta-carotene, and beta-cryptoxanthin. They concluded that other carotenoids, not only the better-known beta-carotene, appeared to be important.

Thus, to summarize these four studies:

- Vitamin E was protective in two studies where it was measured (1, 3).
- Beta-carotene was protective in three studies (2, 3, 4), although in study (4) other carotenes were more protective than was beta-carotene.

Two leaders in this field—Helen Wiseman and Barry Halliwell of King's College, London—have published a review of the mechanisms by which reactive oxygen species promote cancer [234]. They note that "it is increasingly proposed that reactive oxygen species and reactive nitrogen species play a key role in cancer development, especially as evidence is growing that antioxidants may prevent or delay the onset of some types of cancer." Antioxidants may help prevent cancer development by (i) preventing structural damage to DNA; (ii) affecting the mechanism by which cells communicate through chemical signals; (iii) affecting the activity of genes and proteins that respond to stress and that act to regulate the genes involved in cancerous cell growth.

SOME INDIVIDUAL
CANCERS

STOMACH CANCER. Several well-conducted studies have shown that vitamin C supplements have a strong protective effect against cancer of the stomach and esophagus. But recent data suggests that this may have less to do with the antioxidant properties of vitamin C and more to do with the fact that vitamin C can neutralize the potent cancer-producing nitrosamines found in the stomach. Nitrosamines are derived from smoked, cured, or spicy foods, which should therefore either be avoided or taken in small quantities. In four out of five studies, vitamin C supplements reduced the levels of nitrosamines in gastric fluid, reduced the excretion of their metabolic end products in the urine, lowered the ability of the gastric fluid to induce mutations (a precancerous step), and raised the level of activity of DNA repair enzymes that protect against cancer.

Bukin et al. have studied patients with intestinal hyperplasia, a precancerous condition of the stomach [25]. In this condition the stomach tissue contains raised levels of a proto-oncogene ODC (ornithine decarboxylase), which plays a specific and important role in promoting the formation of tumors. The investigators conducted a double-blind, placebo-controlled study of the effect on the tissue level of this proto-oncogene of giving vitamin E (400 mg per day) and beta-carotene (20 mg per day) to patients over a one-year period. Biopsies of the stomach lining showed that vitamin E reduced ODC activity by 65 percent by the end of the year, and beta-carotene reduced ODC activity by 50 percent. In the group receiving placebo, there was no such reduction. Examination of the areas of intestinal metaplasia in the stomach revealed that vitamin E produced significant shrinkage of the metaplastic areas in ten out of fourteen of the patients in one year. Beta-carotene produced significant shrinkage in nine out of eighteen patients. There was no shrinkage of these areas in the patients taking placebo. This study may illustrate an important point, namely that antioxidants are more likely to be effective during the pre-cancerous stage than during the stage after the development of an actual cancer.

BREAST CANCER. Most of the studies in this area have yielded conflicting or negative results. A good review has been provided by Kimmick et al. [104]. Three of special interest warrant description. One study, in which there were 262 cases of breast cancer and 273 normal controls, suggests that the type of breast cancer may be an important factor in determining the response to antioxidant therapy [4]. The cancer patients were divided into two groups—those with negative and those with positive family histories. The intake of vitamins C and E and beta-carotene from dietary histories was estimated, and it was found that in the group with a negative family history, beta-carotene appeared to be protective, whereas vitamins C and E were not. In the group with positive family histories, however, vitamin E seemed to be protective, whereas beta-carotene and vitamin C were not.

Freudenheim et al. studied 297 patients with breast cancer and 311 normal controls [58]. The investigators accepted the notion that a diet rich in fruits and vegetables protects against breast cancer and set out to determine which antioxidants in these foods were important. Their results showed that a diet rich in vitamins C and E, and the carotenes lutein and zeaxanthin as well as fiber derived from fruits and vegetables, reduced the risk of breast cancer, whereas beta-cryptoxanthin, lycopene, and fiber derived from grain did not. The most important factors in reducing risk were the total amount of fruits and vegetables eaten and, among individual items, carotenes, lutein, and zeaxanthin. The authors stressed that all these individual components probably have a team effect and that there are probably relevant factors in fruits and vegetables of which we are still unaware.

Lockwood et al. carried out a two-year survey of 32 patients with breast cancer that had spread to the axillary lymph nodes [127]. They gave the ANICA protocol—a mixture of antioxidants and minerals (including beta-carotene, vitamins C and E, selenium, gamma-linoleic acid, and 90 mg of vitamin Q_{10})—and found that no patients died (the expected number was four); none showed further metastases or weight loss; the need for painkillers was reduced; and six patients showed apparent remission. The authors recommend 300 mg per day of vitamin Q_{10} to attain a blood level of 20 μg/ml. The wide mixture of antioxidants may be a key to their apparent success.

However, other investigations have yielded conflicting or negative results. We can only conclude, on the basis of present evidence, that breast cancer does not respond consistently to dietary manipulation, which confirms the earlier data from epidemiological studies. This may be because breast cancer (as well as prostate cancer) is greatly influenced by hormonal factors. Also, cancer is likely to be a group of related diseases with different pathologies. Clearly, more studies using a wide mixture of antioxidants, like those of Lockwood et al. [127], should be undertaken. This conclusion is supported by a review that states that human studies are "limited but promising" and recommends that antioxidants should always be given as a comprehensive mixture [104].

LUNG CANCER. Ziegler et al. at the National Cancer Institute concur that there is convincing evidence of the protective action of fruits and vegetables against lung cancer but no convincing evidence as to which individual antioxidants may be important [243–245]. However, the same research group in a population-based study of carotenes in smokers found that a diet high in fruits and vegetables was significantly protective against lung cancer, but they concluded that beta-carotene was not the dominant factor [246]. Intakes of alpha-carotene, yellow and orange vegetables, and dark green vegetables (which contain lutein and zeaxanthin) were each more predictive of reduced lung cancer risk than was the intake of beta-carotene. The Netherlands Cohort Study on Diet and Cancer, a prospective study of more than 100,000 Dutch men and women over a 6.3-year period, showed a significant correlation between the amount of vegetables in the diet (excluding potatoes) and a low lung cancer rate [71]. The most protective vegetables were cauliflower, endive, and lettuce. A low consumption of fruit (especially citrus) was associated with increased lung cancer risk.

The complexity of this field is illustrated by two reports published in a 1997 issue of the *American Journal of Epidemiology*. The first, by Yong et al., was the prospective epidemiological NHANES I study of 3,968 men and 6,199 women aged 25 to 74 years with a median follow-up of nineteen years [238]. They measured the intake of antioxidant vitamins during this time and found that vitamin C was highly protective against lung cancer. The quarter of people with the highest intake of vitamin C (computed from an analysis of the diet) had significantly lower rates of lung cancer when compared with the quarter of people with the lowest intake. However, no additional protection seemed to be gained by taking supplements. Vitamin E and carotenes were protective only among people who were current smokers. The authors admit that their finding could be explained by the usual complication that people who took more vitamin C in their diet also were those with generally healthier lifestyles, and that it was perhaps the latter factor that was protective against the development of lung cancer. However, the researchers also found that when vitamins C and E and carotenoids were taken together, the protective effect was much increased. It seems unlikely that people who take all three vitamins would have a much healthier lifestyle as compared with those who take only one antioxidant as a supplement. Thus, it seems more likely that the protective effect is genuine and exemplifies what has been stressed so often in this book—the importance of taking a balanced mixture of antioxidants.

The second report was by Knekt et al. from Finland [108]. They studied 9,959 Finnish men and women aged 15 to 90 who were initially cancer free. A dietary history was obtained with a follow-up period of twenty-four years. During this time 997 of the group developed cancer, and of these 151 developed lung cancer. An analysis of the diet revealed no apparent protective effect by vitamins C and E or beta-carotene, but a very strong protective effect by flavonoids. This protective effect was for lung cancer only. The main source of flavonoids (in this case quercetin) in the Finnish diet is apples and onions. (Finns drink little tea or red wine.) The researchers noted that apples are a poor source of vitamin C and beta-carotene. They also noted that Finns in general had a very low intake of flavonoids as compared with people elsewhere. Oddly, the benefit seemed to be gained from apples and not onions. As we have already seen, Finns, at least southern Finns, seem to react differently than do other populations [5, 174, 175]. Their very high intake of fats and alcohol may be important complicating factors.

Another study on smokers found that adding vitamin E (100–200 mg), with or without vitamin C (500 mg), had no effect on the increased rate of excretion in the urine of a chemical derived from oxidatively damaged DNA [166]. The investigators suggested that cancer-protecting effects of fruits and vegetables (at least in smokers) may result from other anticancer agents, such as flavonoids and polyphenols, in these foods. It is therefore encouraging that companies who make antioxidant supplements have started adding flavonoids and polyphenols to their pills.

COLORECTAL CANCER. Kampman et al. studied the effect of vegetables and animal products on colon cancer risk in Dutch men and women [99]. They found that vegetables reduced the risk in both men and women, whereas red meat increased the risk in women. Other studies on colorectal cancer have failed to achieve consistent results with individual supplements. Some other factor or factors other than antioxidants, such as fiber in a diet high in fruits and vegetables, may account for the reduction of incidence of colorectal cancers.

UTERINE CANCER. Again, the data relating to the value of individual supplements is conflicting. Moreover, lowering serum estrogens, achieved through a vegetarian diet, may be more relevant than antioxidant action on estrogen-related cancers (e.g., breast and uterus). The epidemiological evidence that a diet high in fruits and vegetables appears to be protective against the development of cervical cancer was quoted earlier [17, 43].

ORAL CANCER. A review by Garewal and Diplock states that "all available evidence supports a significant role for antioxidant nutrients in preventing oral cancer" [62]. They recognize that alcohol and tobacco use account for some 75 percent of oral cancer in the West. In epidemiological studies of the role of carotenoids in laryngeal cancer, four out of four were positive; in oropharyngeal cancer nine out of nine were positive. In similar studies of the precancerous condition of leukaplakia, eight out of eight trials yielded positive results. In a clinical trial using daily doses of beta-carotene (30 mg), vitamin C (1 G), and vitamin E (800 mg) in 79 patients with oral cancer, 55 percent reported clinical improvement [101]. Gridley et al. of the National Cancer Institute state that vitamin E protects against oral and pharyngeal cancer [75].

CONCLUSION ON CANCER. There is a vast body of laboratory data, in vitro and in animals, showing a close connection between reactive oxygen species, oxidative stress, and the initiation and progress of cancer. As stated by Franke et al., "evidence for the involvement of endogenously-generated free radicals and oxidants in carcinogenesis and other aging-related diseases is compelling" [57]. Although there is general agreement that a diet low in saturated fats and high in fruits and vegetables is of crucial importance in the prevention of many forms of cancer, there is still a great deal of disagreement as to what the relevant agents in fruits and vegetables are. There is also widespread disagreement about the value of antioxidant supplements. However, through all this murk a general picture arises. Some cancers—oral, upper respiratory tract, lung, and higher gastric tract—seem to be more amenable to prevention by dietary means than others—breast, uterus, prostate, and blood cell. Moreover, dietary mechanisms other than antioxidant action appear to be important in some cases—for example, the antinitrosamine action of vitamin C in the case of stomach cancer and the action of dietary fiber in colorectal cancer. But it remains clear at this time that a diet rich in a variety of fruits and vegetables wins hands down over any vitamin supplements. Such a diet probably provides many anticancer agents besides antioxidants about which we as yet know little. To benefit from the protective effects of dietary chemicals it is best to do what your parents told you—eat your vegetables.

DISEASES OF THE NERVOUS SYSTEM

The brain is particularly vulnerable to oxidative stress for several reasons:

1. It uses oxygen at a faster rate than any other tissue.
2. Its cellular membranes have high levels of easily oxidizable fat.
3. It has naturally low levels of the two antioxidant enzymes CAT and GSHpx.
4. Nerve cells that are killed cannot be replaced, unlike other cells in the body.
5. Several of the brain's chemicals, called neurotransmitters, that are used as signals between nerve cells induce biochemical reactions that lead to the release of reactive oxygen species.

Vitamin C (also known as ascorbic acid or ascorbate) has many functions in the brain unrelated to its vitamin effect in preventing scurvy (which has to do with building up connective tissue in the rest of the body). The brain has very high levels of vitamin C and a specific mechanism for pumping it out of the blood into brain cells. Brain vitamin C plays a role in the synthesis and release of many neurotransmitters, and it also has an important role as the main water-soluble antioxidant in the brain.

The following brain diseases are associated with oxidative stress.

STROKE. A stroke results from the blockage of an artery to the brain. At first the brain suffers from lack of oxygen. This is called the ischemic stage. Then, when the blood reenters the affected region of the brain (called the reperfusion stage), more damage is caused by large quantities of reactive oxygen species that are released. These are primarily responsible for the brain damage suffered by the stroke victim. It is therefore possible that antioxidants, if they can be given soon enough, may prevent some of this damage.

DOWN'S SYNDROME. Down's syndrome is associated with the overproduction of one form of reactive oxygen species—hydrogen peroxide—owing to a genetic fault in the antioxidant enzyme superoxide dismutase (SOD) (see part 1) that makes hydrogen peroxide. These patients have an extra copy of the gene for SOD on chromosome 21. This may seem paradoxical—why should an antioxidant enzyme make too much of an oxidant like hydrogen peroxide? The answer is that the SOD acts on the strongly oxidant superoxide ion and turns it into weakly oxidant hydrogen peroxide. The latter is then detoxified by the second enzyme catalase (CAT), which thus acts in tandem with SOD. In Down's syndrome, the brain level of the enzyme that detoxifies hydrogen peroxide—CAT—is increased in response to the SOD-induced overproduction of hydrogen peroxide. But even with the elevated CAT activity, the overall excess of hydrogen peroxide is harmful to brain tissue.

ALZHEIMER'S DISEASE. The brain in Alzheimer's disease has two neuroanatomical abnormalities visible under the microscope. One type consists of round blobs called plaques, the other of masses of protein filaments called neurofibrillary tangles.

There are two main types of Alzheimer's disease, one with a genetic background and the other without. Some cases of Alzheimer's disease are caused by a defective gene on chromosome 14 that codes for the production of a protein called presinilin-1. This protein forms an integral part of nerve cell membranes. Cells containing this abnormal gene are much more vulnerable than normal cells to oxidative stress.

The brain in Alzheimer's patients shows evidence of oxidative stress, with damage to brain proteins being pronounced [73]. An abnormal, toxic brain protein—beta-amyloid—concentrates in plaques and is toxic to nerve cells by a reactive oxygen species–based mechanism. This mechanism includes activation of the enzyme (called prostaglandin H synthase) that starts the synthesis of inflammation-producing prostaglandins. Levels of this prostaglandin-producing enzyme are greatly raised inside the damaged nerve cells in the disease. In response to this, the levels of the antioxidant enzyme glutathione peroxidase (GSHpx) in the brain is raised to combat the increased oxidative stress. But another antioxidant defense—beta-carotene—gets overwhelmed, so its levels fall. In the cerebral cortex of the brain in Alzheimer's disease there is also evidence of increased fat oxidation. Thus, the brain in Alzheimer's is subject to an intense inflammatory process mediated in part by excess production of reactive oxygen species.

There is now considerable evidence to suggest that nonsteroidal anti-inflammatory agents, such as aspirin or ibuprofen, may prevent or delay the initial onset of symptoms of Alzheimer's disease. Out of sixteen studies on this topic, fifteen had positive results. McGeer et al. conducted an analysis of seventeen epidemiological studies and found a significant correlation between the use of anti-inflammatory drugs and protection against Alzheimer's disease [133, 134]. The discovery of this correlation was made by accident. A series of patients with rheumatoid arthritis were being treated in a long-term study with anti-inflammatory agents. The physicians were surprised to discover that fewer of these patients developed Alzheimer's disease than was expected. Many people take an aspirin a day, as this has been shown to reduce the risk of heart attack. It now seems likely that taking anti-inflammatory agents may also reduce the risk of getting Alzheimer's disease. Anti-inflammatories inhibit the enzyme that makes the inflammation-producing prostaglandins; this reaction is also a potent producer of brain-damaging reactive oxygen species. Thus, it is reasonable to suggest that antioxidants may help here, too, although further studies are needed. In the case of a patient who was unfortunate enough to have both Alzheimer's disease and a form of cancer known as multiple myeloma, the powerful immunosuppressants and anti-inflammatory agents given to him for the treatment of his cancer produced a significant improvement in his Alzheimer's disease for two years, until his death from the cancer [102].

The New England Journal of Medicine reported the results of a double-blind, placebo-controlled, randomized trial of vitamin E (2 G per day), or the drug selegiline (which cuts down the production of reactive oxygen species in the brain), or a combination of the two, in 341 cases of moderate Alzheimer's disease in a mammoth, twenty-three-center study over two years [183]. Practical indicators were measured such as the need for hospitalization, loss of ability to perform the actions needed for daily living, development of severe dementia, and death. The results showed a significant therapeutic effect for both vitamin E and for selegiline in preventing the further development of the illness. However, the combination of the two did not confer any added benefit. In the same issue, Drachman and Leber commented on this paper, pointing out some caveats. They agreed that it is now justified to treat this stage of Alzheimer's disease with either vitamin E or selegiline (but not both). But the study did not address the question as to whether vitamin E would prevent or delay the subsequent development of Alzheimer's disease when given to younger, apparently normal people. Any double-blind, placebo-controlled study designed to answer this question would be enormously difficult and expensive to carry out. This study suggests that vitamin E may be of benefit in cases of established Alzheimer's disease, presumably because of its antioxidant properties. There is now abundant evidence of powerful oxidative stress in the disease. Since there is every scientific indication that antioxidants are more effective when given before the oxidative stress has produced actual cellular damage (see also the section on diabetes), then medical common sense suggests giving people at risk for Alzheimer's disease vitamin E supplements (at around 400–800 mg per day). This level will do them no harm if some elementary precautions, relating to vitamin K and hemorraghic stroke (see part 3), are followed.

PARKINSON'S DISEASE. Parkinson's disease is caused by the destruction of certain nerve cells in a part of the brain called the substantia nigra (or "black substance," so called because these nerve cells contain large amounts of the black pigment neuromelanin). These cells use a chemical called dopamine as their neurotransmitter. Within the cells of the substantia nigra, dopamine is easily oxidized to form a red substance called dopaminochrome, which, in turn, forms neuromelanin. This process releases a large amount of reactive oxygen species. In addition, dopaminochrome is highly toxic to nerve cells. At birth the cells of the substantia nigra contain no neuromelanin; rather, it gradually and steadily accumulates during life. Normally neuromelanin is useful in that it gets rid of the toxic dopaminochrome. It also binds and neutralizes toxic heavy metal atoms such as iron and manganese. However, in excess, it itself becomes toxic and destroys the nerve cell. When a critical mass of the cells in the substantia nigra have been destroyed, Parkinson's disease results. In this disease, there is severe oxidative stress in the substantia nigra with exhaustion of glutathione. (Recall from part 1 that glutathione is the main antioxidant inside the cells of the brain, while vitamin C is the main antioxidant in the watery fluid between the cells.) Hydrogen peroxide also interferes with the storage of dopamine inside the cells in the substantia nigra [48]. Dopamine inside its storage vesicles is protected by antioxidants from oxidation; but dopamine outside these vesicles is more liable to oxidation, because in this locus it has less antioxidant protection. Glutathione scavenges hydrogen peroxide, so, when its level falls, hydrogen peroxide levels rise and dopamine cannot be stored properly. It thus becomes more vulnerable to convert by oxidation to dopamine quinones, poisonous compounds that damage the cells. Because the cells of the substantia nigra contain the major part of the dopamine in the brain and supply it to the rest of the brain, their destruction leads to the symptoms of dopamine depletion. These are tremor, rigidity, and mental impairment—the symptoms of Parkinson's disease.

The current principal treatment of Parkinson's disease is to administer L-DOPA, the precursor of dopamine. In the brain, L-DOPA is taken up and turned into dopamine to replace the dopamine missing in the Parkinsonian brain. However, the beneficial effects of L-DOPA treatment last only a short time: the basic destructive process continues, with the dopamine cells in the substantia nigra being killed off at a steady rate. After about five years the disease progresses to a condition as profound as if L-DOPA had never been administered. Clearly we must develop therapy designed to prevent the killing of these cells and not merely supply by artificial means the product (dopamine) that they make. As with diabetes, for which the main current therapy is insulin, the treatment strategy supplies what is missing but does not address the main problem, which is to prevent the destruction of the insulin-producing beta-cells of the pancreas that caused the disease in the first place.

If Parkinson's disease is due in part to oxidative stress, it would be logical to try antioxidant therapy in an attempt to delay or prevent Parkinson's disease. Moreover, L-DOPA itself is also metabolized in the brain to the toxic compound dopachrome, which can further damage the brain cells. For this reason, Mena et al. advocate vitamin C supplements during L-DOPA therapy [140]. Parkinson's sufferers also have a higher incidence of heart disease than do normals, possibly because of the same raised level of oxidative stress.

People with Parkinson's disease typically have normal blood levels of vitamins A, C, and E, and Scheider et al. have claimed that there is no correlation between the disease and the previous intake of antioxidant vitamins [188]. A contrary result was obtained by de Rijk et al. [41] in a study of 5,342 subjects aged 55 to 95. A semiquantitative food-intake questionnaire was given from 1990 to 1993. Thirty-one subjects subsequently developed Parkinson's disease; they had a significantly lower intake of vitamin E, but not vitamin C or flavonoids, than did those who did not develop Parkinson's disease. Another such study in Hawaii produced the same result.

There have also been some studies of treatment of cases of established Parkinson's disease by antioxidants, but the results have been inconclusive. However, the damage to the pigmented neurons of the substantia nigra starts at least five years before the earliest clinical symptoms of Parkinson's disease show themselves. Therefore, giving antioxidants to patients already with symptoms may be a case of shutting the stable door after the horse has gone. As there is at present no way to identify which people will develop Parkinson's disease five years later, the only way to apply this strategy would be to raise the antioxidant intake for everyone over the age of 40.

MOTOR NEURON DISEASE. Motor neuron disease goes by several names. Many know it as Lou Gehrig's disease; the medical term is amytrophic lateral sclerosis, or ALS. The physicist Stephen Hawking suffers from ALS and has been a long-term survivor of the disease. Most patients die within three to five years of diagnosis. It is caused by a selective death of the large motor nerve cells in the spinal cord. Without the transmission of motor signals to the muscles, widespread paralysis results. There are two forms of the disease, genetic and sporadic or nongenetic. In the former type, the gene at fault is the one for making the antioxidant enzyme SOD and is found on chromosome 21. The brain in these cases shows lower antioxidant defenses and chemical signs of increased fat oxidation.

Reider and Paulson tell the interesting story of how vitamin E was used some fifty years ago to try to treat ALS [173]. The patients included the famous baseball player Lou Gehrig, alas without benefit. However, these researchers suggest that, now that some cases of the disease have been traced to a genetically abnormal antioxidant enzyme (SOD), these may represent a subtype of the disease and that further trials are indicated.

HUNTINGTON'S DISEASE. Huntington's disease, also known as Huntington's chorea, is characterized by profound dementia, abnormal emotional reactions, disturbed behavior, and involuntary movements called chorea. It is caused by a mutation in a dominant gene and is characterized by severe nerve-cell loss in the basal ganglia and parts of the cerebral cortex. The basal ganglia are structures in the brain that control the programming of voluntary movements. Within the affected brain tissues, the enzymes in the mitochondria that provide the energy the cell needs are severely affected. DNA removed from an affected part of the brain of a patient with Huntington's disease shows an abnormal chemical change that was brought about by attack by a reactive oxygen species [23].

SCHIZOPHRENIA. This is a severe illness that affects one in every hundred people. It is characterized by hallucinations, delusions, paranoia, and emotional disturbances. It used to be thought of as a psychological illness, but we now know that it is an organic disease of the brain [202]. There are two types of the illness. Type I shows acute "positive" symptoms such as hallucinations and excitement. It has a strong genetic component, responds well to treatment with antipsychotic drugs, and the brain shows no signs of physical degeneration on a scan. Type II is characterized by "negative" symptoms such as social withdrawal and apathy and responds poorly to most medicines except clozapine, an atypical antipsychotic drug. In type II cases there is no genetic component but the brain typically shows signs of physical damage on a scan.

Modern techniques have shown at the microscopic level that the brain in type II schizophrenia has been damaged. In many brain regions there are fewer nerve cells than normal; in a few areas there appear to be too many; and, in other key areas, there is a loss of the interconnections between cells. Because the job of the brain is conducted mainly by neurons sending each other signals, loss of part of the network that carries these signals can have serious consequences.

In this system, if the number of connections between neurons is reduced below a certain level, neural computations cannot proceed normally, and phenomena called parasitic foci develop in which a certain computation enters into a vicious repetitive cycle that cannot be changed by external influences; in the human brain affected by schizophrenia, delusions and hallucinations result.

How are these key connections lost in schizophrenia? In order to understand the mechanism, we must discuss some specifics of what happens at these connections between neurons. The gap between the end of one neural process (or axon) and the next neuron is called a synapse. The chemical that carries the message from one neuron to the next across this gap is called a neurotransmitter. Neurotransmitters can either be excitatory or inhibitory. Excitatory neurotransmitters, when released across the synapse, cause the second neuron to fire. Inhibitory neurotransmitters, when released, cause it to stop firing. In the brain in type II schizophrenia there is a 50 percent loss of excitatory synapses.

The main excitatory neurotransmitter in the human brain is a chemical called glutamate, more familiar as the monosodium glutamate used in much Asia Glutamate is a very toxic molecule, and it is remark- it for such a key role in neurotransmission. Yet ion. Glutamate exerts its toxic action on nerve ctive oxygen species it generates in the second This process is one of the ways in which the et rid of unwanted synapses. Synapses, even in constant state of flux. New synapses are contin- d ones deleted. We are concerned here with the synapse deletion [201].

ns where glutamate is the neurotransmitter, there rs. One of these is called the NMDA receptor. or results in an inflow of calcium ions into the cascade leading to activation of two enzymes. din H synthase (or PG H synthase), which we zyme that initiates prostaglandin synthesis. The se leads to the release of large quantities of toxic ncluding hydrogen peroxide. Hydrogen perox- is not very reactive but it can diffuse widely in and out of cells.

Under certain conditions, especially in the presence of free iron, it can form more toxic reactive oxygen species such as the hydroxyl radical.

The other relevant effect of the NMDA receptor is to activate another enzyme called nitric oxide synthase, which makes nitric oxide (NO). Nitric oxide is a gas capable of diffusing widely from its point of origin. There is evidence to show that both these compounds—hydrogen peroxide and nitric oxide—have important functions in the body as signaling molecules. Moreover, they both have pro-oxidant properties, and, being freely diffusible molecules, they can enter the glutamate synapse itself. Here they can actually eliminate the synapse. One important factor that determines whether a synapse is maintained or deleted is the balance in the synapse between neurotoxic oxidants (hydrogen peroxide and nitric oxide), which will tend to delete the synapse, and neuroprotective antioxidants, which will tend to maintain it.

The antioxidant defenses at the synapse consist of three main components. The first is vitamin C, which is released by the axon terminal during the process in which the action of glutamate is terminated by pumping it back into the synaptic terminal, where it can be recycled to be used again. Thus, the brain very neatly arranges for a flood of protective vitamin C to be pumped into the synapse just when the damaging hydrogen peroxide is diffusing back from the postsynaptic site. The second is an antioxidant—carnosine—which is found in glutamate nerve terminals and released together with the glutamate. The third is the most interesting. Attached to the side of most glutamate synapses is a dopamine terminal, which is triggered when any activity of the organism is rewarded by what we call positive reinforcement—that is, food or some other pleasurable stimulus. Some of the dopamine it releases, whenever such a stimulus is received, enters the adjacent glutamate synapse. Dopamine happens to be a powerful antioxidant—a key fact discovered by two scientists in Japan, Jiankang Liu and Akitane Mori [125]. Dopamine here will increase the antioxidant status of the synapse and tilt the balance of synapse growth and deletion in the direction of the former. This may be one of the biochemical ways in which conditioned reflexes are built up and learning takes place: it tends to strengthen those behaviors that lead to positive reinforcement, because the circuits that underlie that behavior are strengthened by growing more synapses. Likewise, if less dopamine is released, the balance in the synapse tilts the other way, and the probability increases that that synapse will be pruned.

However, using dopamine as an antioxidant in this way carries a risk. For, in the absence of sufficient levels of other antioxidants in the synapse, dopamine easily converts spontaneously (by oxidation) to toxic quinones, including dopaminochrome. Furthermore, under these conditions of lowered antioxidant defenses, nitric oxide will also convert to poisonous compounds called reactive nitrogen species (such as peroxynitrite), which will attack dopamine and hasten its conversion to the toxic quinones. These reactive nitrogen species and dopamine quinones will attack the synapse and can destroy it.

Evidence suggests that it may be this system that goes wrong in type II schizophrenia:

1. Antioxidants mop up or otherwise inactivate the reactive oxygen species that would otherwise delete the synapse. Antioxidants also slow down the production of quinones from dopamine. These antioxidant mechanisms, particularly those relating to vitamin C and to GSHpx, are faulty in schizophrenia.

2. The black pigment in the brain, neuromelanin, which is made from the toxic quinones, is a means of taking them out of circulation. There is preliminary evidence that neuromelanin is abnormal in some types of schizophrenia.

3. Some brain dopamine is normally turned into a compound called 5-cysteinyl dopamine, which is in itself a good antioxidant. This process also prevents the formation of the toxic quinones. Levels of this compound are elevated in the schizophrenic brain. However, this compound can itself lead to the production in the brain of neurotoxic derivatives (called benzothiazines) that have been implicated in the cause of Parkinson's disease.

4. Another defense against the formation of toxic quinones from dopamine a process called transmethylation, converts them into harmless compounds. In this process a methyl group (composed of one atom of carbon and three of hydrogen) is attached to an active site on the molecule and renders it nontoxic. My work with Leland Tolbert, John Kelso, and others has shown that this reaction is also weak in schizophrenia.

Direct chemical evidence now exists that the quinones derived from dopamine are present in the human brain. Indeed, they must be, because they lie on the chemical synthetic pathway by which neuromelanin is made, and neuromelanin is certainly present in the brain.

Epinephrine and dopamine are closely chemically related—in fact, epinephrine is made out of dopamine. In 1954 Hoffer, Osmond, and I discovered that the quinone derivative of epinephrine, adrenochrome, when given to normal people produces in them many of the symptoms of schizophrenia [89]. This experiment was confirmed by three other groups of scientists, one in the United States, one in Germany, and one in Czechoslovakia. Epinephrine is used in the brain as a neurotransmitter by only one small set of nerve cells. It is more likely that close relatives of adrenochrome—the quinones derived from dopamine and norepinephrine—are involved in the schizophrenic process. Norepinephrine is found in a nucleus of the brain called the locus coeruleus; the nerve cells here also contain neuromelanin, indicating that norepinephrine, too, is metabolized in the brain by the quinone pathway. There is evidence that the nerve cells of the norepinephrine pathway are also damaged in schizophrenia, and this damage may be similarly affected by reactive oxygen species and the quinone derivative of norepinephrine. No one has as yet tested the effect of these quinones derived from dopamine and norepinephrine given to normal people, but it may well be that they act as hallucinogens, like their close relative adrenochrome, the quinone derived from epinephrine.

Thus, to summarize, reactive oxygen species and oxidized derivatives of brain dopamine called quinones may be involved in schizophrenia, on the evidence that they are agents that may prune synapses and in schizophrenia defenses against them (antioxidants, O-methylation, and neuromelanin formation) are faulty. It is therefore worth investigating whether boosting antioxidant defenses and/or cutting down reactive oxygen species production is of benefit in the illness.

People with schizophrenia are low in antioxidant defenses. Plasma levels of vitamin C are low, and more vitamin C than normal is needed to raise blood levels to a given level, suggesting an increased usage in combating oxidative stress [211]. Glutathione levels in blood are also low, and there is a strong correlation between low levels of the antioxidant enzyme GSHpx and the raised levels of brain damage in schizophrenia.

My work with Hoffer and Osmond has demonstrated that schizophrenia may also be associated with a diminished capacity for dealing with toxic metabolites in the brain produced from the oxidation of epinephrine, dopamine and norepinephrine. Many antischizophrenic drugs, such as chlorpromazine, are powerful moppers-up of the toxic hydroxyl radical and inhibitors of fat oxidation. The potent antischizophrenic drug halo-peridol acts differently: it mops up another kind of reactive oxygen species called hypochlorous acid. Thus, it is possible that the beneficial effects of these drugs in schizophrenia may be due in part to their anti-oxidant properties. Reports have indicated that antioxidants such as vita-min C benefit schizophrenic patients [100]. More work in this area seems to be called for.

PSYCHOLOGICAL
STRESS
Some experiments with rats have recently shown that severe psychological stress in rats (induced by restraining the animal for eight hours, which a rat finds exceedingly stressful) leads to severe oxidative stress. The oxidative stress is charac-terized by exhaustion of antioxidant defenses and an elevated level of oxi-dized fats. The principal damaging agent in this type of stress is the hydroxyl radical that produces stomach ulcers in rats under these cir-cumstances [39]. Such ulcers may be prevented by antioxidants. There are as yet no human experiments designed to investigate the possible rela-tion between psychological stress and oxidative stress, but the results of such a trial would be interesting.

DIABETES
This disease is caused by a defect in sugar metabolism owing to a lack of insulin. Even with insulin treatment, it still remains a major cause of blindness, kidney failure, disorders of the heart, brain, and vascular systems, and limb amputations. Type I diabetes, also called insulin-dependent diabetes, is characterized by marked lack of insulin, its early age of onset, its severity, and the need to take insulin injections. Insulin is produced by the beta-cells of the pancreas. Type I diabetes is an autoimmune disease in which the body's immune system

mistakes its own beta-cells in the pancreas for foreign intruders and proceeds to destroy them by a mechanism that includes an attack by reactive oxygen species. Beta-cells normally have a low level of antioxidant protection. Levels of reactive oxygen species are also generally increased—that is, not only in the pancreas—because the diabetic's white cells produce more reactive oxygen species and more prostaglandins are synthesized. (You will remember from part 1 that the process of making prostaglandins releases large amounts of reactive oxygen species.) Moreover, two mechanisms that involve the abnormal metabolism of blood sugar—glucose—also yield reactive oxygen species as a by-product. There is also evidence that free iron and copper ions are released from the stores where they are usually kept from doing harm. Free iron and copper circulating in the body fluids are extremely toxic oxidants. Thus, in diabetes the increased levels of free iron and copper ions adds to the level of oxidative stress.

Animal studies show that antioxidants can prevent or alleviate various forms of experimentally induced diabetes. In diabetic rats the synthetic antioxidant probucol protects the heart against oxidative stress. In human diabetes, blood levels of markers for oxidative stress are increased and levels of protective antioxidants decreased (in particular uric acid and vitamin C) [133]. Vitamins C and E inhibit fat oxidation in diabetics. Non-insulin-dependent diabetics have significantly lowered levels of vitamin C in the blood [133]. Tests of vitamin C in human diabetics have shown that it improves vascular performance, leading to a fall in blood levels of LDL (bad) cholesterol and a rise in protective glutathione levels [160]. Diabetics also have been reported to have lowered blood levels of vitamin A, glutathione peroxidase, and uric acid, as well as having signs of oxidative damage to the DNA in blood cells [78]. In other tests nicotinamide (vitamin B_3) at a dosage of 3 G per day reduced the need to give insulin to diabetics. Vitamin B_3 is one of the helpers of the antioxidant action of vitamin E as described in part 1. Santini et al. conclude from their studies that in insulin-dependent diabetes oxidative stress is increased and antioxidant defenses are defective "regardless of duration, metabolic control or presence of complications" [184].

Chase et al. reported an initial negative result in a preliminary test of nicotinamide in diabetic children and adolescents [33]. Though this was disappointing, they reasoned that by the time actual diabetes develops, the damage to the pancreatic beta-cells has already been done. Therefore, they suggested that antioxidant therapy be started in the pre-diabetic stage. They tested this theory in a further study of twenty-two children at high risk for developing diabetes . These children had a first-degree relative with diabetes, as well as antibodies in their blood against beta-cells and impaired insulin release. They gave eight children placebo and fourteen children 3 G per day of nicotinamide. All eight of the children on placebo developed diabetes by the end of the study, but only one of the children on nicotinamide did. The authors felt that this result was most promising and again stressed that the nicotinamide must be given before the beta-cells are destroyed.

The synthetic antioxidant silymarin has been tested in diabetic patients who had cirrhosis of the liver [223]. Thirty patients were given insulin only and the other thirty were given insulin plus silymarin. After twelve months the patients who had had the silymarin showed significant clinical improvement as compared to the controls. Indices used included fasting blood glucose levels, glycosuria, fasting insulin levels, insulin requirements, and MDA levels.

In a study of eighty diabetics with damage to the retina and peripheral nerves, twenty patients were given the antioxidant lipoic acid, twenty received vitamin E, twenty were given selenium, and twenty received placebo [156]. The first two groups showed a significant improvement, as measured by less fat oxidation, better leg sensation, improved reflexes, and less albumin in the urine. The antioxidant lipoic acid is currently being used in the treatment of diabetic nerve damage by Dr. Packer's group. Lipoic acid crosses the blood-brain barrier, which N-acetyl cysteine does not. A double-blind, placebo-controlled study has been carried out on the effects of lipoic acid on a disorder of heart function (cardiac autonomic neuropathy) that occurs as a complication of diabetes [241]. Thirty-nine patients received 800 mg per day of lipoic acid and thirty-four others received placebo. The investigators reported a modest but significant improvement in heart function in the group that received the lipoic acid when compared with those that received the placebo. There were no side effects noted. Another double-blind, placebo-controlled trial of the antioxidant amino acid L-arginine in human diabetes found that it strongly reduced the levels of fat oxidation. The complex interactions between fat intake and antioxidant status in diabetics has been illustrated by a study in which a simple change in diet (less fat, more carbohydrate) led to an increase in blood levels of vitamins C and E [6]. This is because the lower fat intake produced less oxidative stress, thus sparing the antioxidant vitamins.

Pregnant diabetic rats fed either the synthetic antioxidant BHT (butylated hydroxy toluene) or vitamin E had a lower incidence of offspring with congenital abnormalities than did untreated diabetic rats. This was confirmed by a second study in rats using a cocktail of vitamin E (400 mg per day) and two compounds reported to be deficient in diabetes—arachidonic acid and myo-inositol [172]. The untreated diabetic rats had a high incidence of abnormal offspring, including a 23.7 percent rate of neural tube defects, as compared with 4 percent of normal rats. The treatment reduced the incidence of neural tube defects in the diabetic rats to normal levels. These abnormalities are thought to be due to high blood-sugar levels in diabetes that may interfere with the transport of the key antioxidant glutathione in the body. In humans, the birth defects associated with diabetes can be prevented by a strict adherence to a treatment and dietary regimen designed to limit raised blood-sugar levels [171]. Antioxidant supplements, however, may make this treatment more effective. No human work has as yet been done to test this hypothesis. The conclusion is that antioxidant therapy seems promising in diabetes, especially if the prediabetic stage can be diagnosed and treated.

EYE
DISEASES

CATARACTS.   The transparent crystalline proteins of the eye, unlike most proteins, are not continually replaced. They are also at risk for oxidative damage from ultraviolet light. In animal experiments vitamins C and E and some synthetic antioxidants protected the lens against oxidative stress. The importance of vitamin C is indicated by the fact that there is an active transport system (pump) for it in the watery humor and the lens. Several epidemiological studies have suggested a protective role for antioxidants against cataract formation.

(1) Hankinson et al. carried out a study of more than 50,000 nurses for an eight-year follow-up period [77]. They measured the intake of antioxidants in the diet and compared the lowest intake group with the highest intake group for carotenes and found a 39 percent reduction in cataracts in the latter. Vitamins C and E and riboflavin in the diet were ineffective. Spinach rather than carrots was the most effective vegetable. Dietary vitamin C had no effect nor did multivitamin tablets. However, vitamin C supplements (average dose of 50–500 mg per day) taken for ten or more years led to a 45 percent reduction in cataracts. The authors quote three previous studies that obtained the same result and one that failed to do so. Why did the study by Hankinson et al. find that dietary vitamin C was not protective whereas vitamin C supplements were? Probably because of the higher dose in the case of supplements.

(2) In a carefully controlled trial involving 152 subjects and 152 controls, vitamin C supplements (300–600 mg per day) over five years appeared to reduce the cataract rate by 70 percent [178]. Vitamin E supplements (400 mg per day) for five years appeared to reduce the rate by 55 percent.

(3) A study of 1,380 ophthalmic outpatients aged forty to seventy-nine indicated that the regular use of vitamin supplements significantly lowered the incidence of cataract [121].

(4) A fifteen-year follow-up study compared forty-seven patients with cataracts and ninety-four matched controls [106]. Low serum levels of beta-carotene and vitamin E correlated with more cataract formation. There was a threefold difference in the incidence of cataracts between the one-third of subjects who had the highest serum levels and the one-third who had the lowest, indicating that these antioxidants protected against cataract formation. Levels of vitamin C were not measured.

(5) Jacques et al., in a study of 247 females aged 56 to 71, showed that vitamin C supplements reduced the incidence of early lens opacities by 77 percent and of moderate lens opacities by 83 percent. However, the vitamin C had to be given for a minimum period of ten years for the benefit to be obtained. Shorter periods showed no effect. The researchers conclude that "long-term consumption of vitamin C

supplements may substantially reduce the development of age-related lens opacities" [97].

These studies can be summarized as follows:

- Vitamin E appeared to be protective in two studies (2 and 4).
- Vitamin C appeared to be protective in six studies and not in one (1, 2, 5, and those quoted by 1).
- Carotenes appeared to be protective in two studies (1 and 4).
- Vitamin supplements appeared to be protective in one study (3).

However, when one takes into account that there are three types of cataract—nuclear sclerosis, cortical opacities, and postsubcapsular—things get more complicated. There have not been enough studies as yet to determine reliably which subtypes respond or do not respond to antioxidant therapy.

The reaction of the experts to all this information has been mixed. Christen concludes that the epidemiological data are "wildly conflicting" with many uncontrolled confounding variables [34]. Gershoff says they are "promising but unproven" [63]. Taylor believes the consensus is impressive and suggests that the optimum levels of nutrients for the delay of cataract should be defined as soon as possible [214]. Optimum levels for vitamin E will, he suggests, probably require supplements. Diplock claims that "the laboratory and epidemiological evidence is now so strong that it is essential to ensure that the population is adequately supplied with E, C and beta-carotene" [43].

MACULAR DEGENERATION. In the case of advanced age-related macular degeneration (a disease in which the center of the retina degenerates), one large multicenter study of 356 patients and 520 normal controls has shown that a higher dietary intake of carotenoids was correlated with a 43 percent reduction of macular degeneration, but vitamins C and E were ineffective [192]. The particular carotenoids most likely to be involved were not beta-carotene but lutein and zeaxanthin, which are found in dark green leafy vegetables such as spinach, collard greens, kale, and turnip tops; carrots are very low in these two compounds. However, this result may have little to do with antioxidants. Lutein and zeaxanthin happen to be the two dominant yellow pigments in the macula (center) of the retina that filter out the damaging blue light. Beta-carotene and lycopene (from tomatoes) are absent from the macula. Increasing the serum levels of lutein, by dietary manipulation, leads to increased lutein in the human eye. Thus lutein and zeaxanthine may act to protect the retina directly by their function as natural pigments against light damage, rather than as general antioxidants. However, another study of 167 cases of macular degeneration implicated low levels of lycopene rather than lutein and zeaxanthin in the disease [131]. Red blood cells in cases of macular degeneration have been reported to have significant decreases in levels of SOD and glutathione peroxidase [148]. Landrum et al. suggest that long-term supplementation for individuals having low levels of macular pigment could be of benefit [115]. A diet rich in dark green leafy vegetables and/or supplements with lutein and zeaxanthin, and with tomatoes and/or supplements of lycopene, may help protect against this form of blindness.

RESPIRATORY

DISEASES   The lung is at risk of oxidative stress because of the large amounts of oxygen that pass through it when we breathe. Oxidative stress in the lung leads to damage to the cell membranes, inflammation, and bronchoconstriction.

ASTHMA. Asthma is a chronic respiratory disorder that results in constriction of the bronchi and difficulty in breathing. It is associated with oxidative stress because the white cells in the epithelial lining of the bronchi produce more reactive oxygen species than normal. Risk factors for asthma include low dietary intake of vitamins C and E and selenium, high body iron, and exposure to environmental toxins such as lead, which poisons several enzymes in the body on the pathway making the important antioxidant glutathione. It has been shown in eight out of ten studies that the children of smokers have an increased incidence of asthma. Oxidant exposure in infancy may lead to asthma in later life. Asthmatics have low blood levels of the antioxidant enzyme GSHpx and low levels of the trace mineral selenium, which is a part of this enzyme complex. This leads to increased oxidative stress from reactive oxygen species like hydrogen peroxide, which are normally neutralized by selenium. New Zealand has low levels of selenium in the soil, and New Zealanders have a high incidence of asthma and a high mortality from the disease.

A large epidemiological survey confirmed that a low dietary intake of vitamin C is correlated with asthma and that the symptoms of asthma are made worse by environmental oxidants and better by vitamin C [79]. Vitamin C is the major antioxidant in the fluid that covers the surface of the lung where it protects against environmental oxidants, including toxic nitrogen oxides in smog. Antioxidant defenses are particularly low during an acute asthmatic attack. The current evidence suggests that anti-oxidants should protect against asthma and alleviate its symptoms. Interestingly, beta-blockers (such as isoproterenol), which are widely used in asthma, are also potent antioxidants; the authors of one study suggest that this may contribute to their therapeutic effect in asthma by mopping up the excess reactive oxygen species found in the disease [68].

A study carried out in rural China found that increasing the intake of vitamin C by 100 mg per day improved lung function as measured by forced expiratory tests [92]. The improvement was estimated to equal one year of aging-related loss of lung function. A study of 393 nonsmokers found that blood levels of vitamin C were inversely related levels of a chemical marker for fat oxidation [190]. This suggested that vitamin C was protective against fat oxidation. A recent double-blind, placebo-controlled study of seventeen asthmatic adults showed that supplementation with vitamin E (400 mg per day) and vitamin C (500 mg per day) led to a reduction in their sensitivity to ozone that was most marked in the six most severe cases [223].

ACUTE RESPIRATORY DISTRESS SYNDROME (ARDS).   In ARDS the anti-oxidant defenses in the blood are lowered, and there is a fall in the protective elements of the plasma. In one study thirty-two patients with ARDS were treated, half with conventional treatment and half with antioxidants (vitamins C and E, selenium, and N-acetyl cysteine) [186]. The mortality rate in the first group over four weeks was 71 percent, in the second group 37 percent—a very significant difference. In a second clinical study the synthetic antioxidant EUK-8 was found significantly to protect against the type of acute lung injury that occurs during shock. In a double-blind, placebo-controlled prospective clinical trial involving five intensive care units in the United States and Canada, ARDS patients were given the antioxidants N-acetyl cysteine and procysteine [13]. There was no effect on overall mortality (possibly because the sample size was small), but there was a significant improvement in lung and heart function.

RESULT OF CIGARETTE SMOKING. Another lung condition in which reactive oxygen species play a role is the result of exposure to cigarette smoke—both active smoking and passive exposure to smoking. Cigarette smoking is the single largest cause of premature death in industrialized societies and of course affects many other systems besides the lungs. Tobacco smoke has high levels of reactive oxygen species and pro-oxidant oxides of nitrogen, as well as a variety of cancer-producing chemicals. Every puff of cigarette smoke inhaled contains 10^{15} (1 followed by 15 zeros) molecules of reactive oxygen species. In smokers indices of oxidative stress are raised, whereas the levels of vitamin C and the antioxidant enzyme CAT are reduced in the fluid lining their lungs. Glutathione levels are reduced in the lung cells themselves and plasma levels of vitamin C (but not vitamin E) are also reduced. Smokers also excrete in the breath excessive amounts of the gas ethane, which is a product of fat oxidation. The more cigarettes they smoke the more ethane they excrete. Vitamin E reduces fat oxidation in both smokers and nonsmokers. In one three-week study ten smokers were given an antioxidant cocktail that included 6 mg of beta-carotene, 200 mg of vitamin E, and 250 mg of vitamin C [8]. There resulted a 29 percent reduction in the amount of ethane excreted. Smokers excrete four to ten times the normal amount of 8-OHαG (a measure of oxidative attack on DNA) than normals.

Brown et al. studied fifty smokers and fifty normal controls in a double-blind, crossover trial of vitamin E (280 mg) versus placebo [22]. They measured the sensitivity of the red blood cells to oxidative stress induced by the pro-oxidant hydrogen peroxide, as well as levels of key antioxidant enzymes. They found that vitamin E protected the red blood cells against oxidative stress in both smokers and nonsmokers and raised the levels of some antioxidant enzymes. Lykkesfeldt et al., using new and accurate methods, measured the amount of vitamin C and oxidized vitamin C in the blood of smokers [130]. They found that levels of ordinary vitamin C were below normal, whereas the levels of the oxidized form of vitamin C were much above (18 times) normal. When vitamin C recycles vitamin E, the vitamin C is turned into oxidized vitamin C (see part 1). This was evidence of severe oxidative stress and depletion of vitamin C. They recommended that all smokers needed to maintain their blood levels of vitamin C at about 70 μl/L. This would require an intake of at least 200 mg per day of vitamin C. Lehr et al. say that there is now enough experimental and epidemiological evidence to warrant clinical trials of vitamin C in diseases associated with cigarette smoking, such as pulmonary emphysema, chronic obstructive pulmonary disease, and atherosclerosis [120]. It would seem logical therefore that smokers who cannot stop smoking should increase their antioxidant defenses.

EXPOSURE TO SMOG. Nitrogen dioxide (NO_2) is a prominent and powerful oxidant present in smog. In human plasma exposed to NO_2, there is a rapid fall in levels of vitamins C and Q_{10} and a slower fall in vitamin E, bilirubin, and protective protein groups. Humans exposed to excessive levels of NO_2 (such as glass craftsmen and braziers) show greatly excessive levels of fat oxidation (up to one hundred times the normal) and are in special need of antioxidant protection. The same may be true of people who live in smog-laden cities such as Los Angeles and Mexico City, but specific studies need to be carried out to determine this relationship.

Diesel oil fumes contain as much as one hundred times as much toxic nitrogen oxides as do gasoline fumes. If blood plasma is exposed to diesel oil fumes, there is a profound fall in vitamin C and urate levels and a lesser fall in vitamin E levels. The blood proteins also show signs of oxidative damage.

AGING   Aging is not a disease but a normal life process that involves physical and biochemical changes in the body. It is generally agreed that the only reliable way to increase life span through diet is to eat less (known in the field as caloric restriction). This insight is related to a correlation between increased food consumption and oxidative stress. A study of rats fed as much as they wanted and rats given a restricted diet showed that the increased calories in the diet led to raised measures of fat oxidation in the blood. Caloric restriction lowers reactive oxygen species production and levels of protein and DNA oxidation. It also reduces the normal fall with age of the antioxidant enzyme CAT in blood. The mechanism for this effect is unknown, but it may involve the enzymes and small molecules responsible for the removal of reactive oxygen species.

There is some evidence to suggest that increased oxidative stress and a decline in antioxidant defenses play a role in aging. There are reports of high levels of reactive oxygen species and decreased antioxidant activity in elderly people, particularly in the adrenal glands and in the brain [181]. The aging brain contains higher than normal levels of oxidized fats. Moreover, the aged have impaired immune responses that renders them more susceptible to infections like influenza and pneumonia. Increased intake of some antioxidant nutrients—especially vitamin E—improves the immune response [141].

Leewenburgh et al. claim that aging is associated with a fall in both enzymatic and small-molecule antioxidant defenses in many important organs (liver, brain, heart, kidney, but not muscle) [118]. Antioxidant defenses in the adrenal gland decrease with age, which, in turn, leads to a decline in the production of adrenal cortical hormones [7]. The adrenals of young animals are well protected against oxidative damage, and have high levels of several antioxidants, including vitamins C and E, and show low levels of fat oxidation. A diet deficient in vitamin E leads to increased fat oxidation in the adrenals and decreased production of adrenal hormones. A recent study in Holland found that beta-carotene appeared to protect against age-related loss of cognitive function. However, Cals et al. have concluded, following a study in Paris, that aging by itself does not lead to oxidative stress as long as good general health and good nutrition are maintained [27].

There is as yet no evidence to show whether antioxidant supplements will actually slow the aging process in humans. Further studies are needed. But this remains a possible benefit to be hoped for by people who attend conscientiously to their antioxidant intake. However, whether increased oxidative stress, weakened immune function, and decreased antioxidant defenses are the hallmarks of normal aging or are due to the impaired nutrition common among many elderly people is still an open question.

HIV INFECTION

AND AIDS   AIDS patients are under powerful oxidative stress from two sources. One is the excessive production of reactive oxygen species by their white blood cells. The other is the excessive production of certain proteins, called cytokines, that control the immune system. Many cytokines also stimulate processes that result in oxidative stress. These patients show deficiencies of zinc and selenium (two dietary metals that are essential for the proper function of antioxidant enzymes) and of the antioxidant glutathione. As the disease progresses, reactive oxygen species production and levels of fat oxidation increase as a result of the attack of the HIV virus on the immune system. This results in a progressive lowering of antioxidant defenses, in particular blood levels of glutathione, zinc, selenium, vitamin E, carotenoids, and GSHpx, which are exhausted by the high level of continuing oxidative stress. AIDS patients show excess fat oxidation. Levels of fat oxidation start to rise early in the disease before any symptoms develop.

Some experts recommend that treatment of AIDS should include therapies aimed at restoring depleted glutathione levels. One such agent is NAC, which has been safely used in medicine for twenty-five years. The only problem is that NAC is incompatible with trypsin, chymotrypsin, and many antibiotics. There should be a gap of at least two hours between taking NAC and any of these. It has also been suggested that NAC may exert part of its reported therapeutic action by raising blood cystine levels, which are low in the disease. These low levels lead to complications such as muscle wasting and reduced immune responses [46, 47]. Other promising drugs of the same type are alpha lipoic acid and glutamine, which the body uses to make glutathione.

A contrary opinion is expressed by Aillet et al. of the Institut Pasteur [2]. Following test-tube (in vitro) studies of some antioxidants, including NAC, on the replication (growth and spread) of the HIV virus, the investigators found that the antioxidants were only partly successful in blocking HIV multiplication in one type of white blood cell and unsuccessful in other types. Moreover, they found that the high doses of antioxidants needed to produce this moderate effect caused an unwanted blockade of other important blood cells called monocytes. They concluded that these antioxidants cannot counter the intense activity of the multiplication of the HIV virus but may aggravate immune disturbances in HIV patients by the monocyte blockade. Moreover, in HIV patients low doses of antioxidants may have a paradoxical effect of increasing HIV multiplication. This is unwelcome news for the many AIDS patients who currently take large amounts of antioxidants such as NAC.

The effects of beta-carotene and selenium supplementation have been tested in patients infected with HIV. Although measures of oxidative stress were reduced, it was disappointing that there was no positive clinical effect noted. Everall et al. have found that vitamin C slows the growth of the HIV virus in test-tube experiments [54]. Moreover, the level of vitamin C is lowered in the brain of patients who died of AIDS. The investigators suggest that vitamin C has two roles in AIDS: the first is to raise the level of the antioxidant defenses, and the second may be to kill the HIV virus.

AIDS patients with low levels of vitamins A, beta-carotene, B_6, and B_{12} have a poor prognosis, and supplements of these vitamins increase the survival rate [212]. Another study has shown that HIV-infected patients have a lower risk of developing actual AIDS if they maintain a high intake of vitamin E [1]. However, a high intake of zinc increases the death rate. In the case of beta-carotene the optimum amount is about 10 mg per day. Too much is harmful. These results may be due to the effect of the B vitamins on the immune system.

Recently Dröge et al. suggested that AIDS patients must be treated on an individual basis, adjusting the dose to the needs of the patient as is done in the case of insulin treatment of diabetes [46]. Blood levels of cystine, cysteine, and glutamine need to be monitored. This creates difficulties for attempts to carry out double-blind, placebo-controlled therapeutic trials using only one dose of NAC. Clearly, much more research needs to be done in this field.

SHOCK   Shock is an acute medical emergency caused by injury, blood loss, severe burns, infection, and so on. Septic shock, due to a bacterial infection, is the leading cause of death in intensive care units. The bacterium releases a toxin that provokes an overproduction of cytokines and reactive oxygen species. Septic shock patients have increased fat oxidation rates and exhausted antioxidant defenses. In animal experiments antioxidants (vitamin E, NAC, and SOD—if this is given before the sepsis starts) have proven to offer effective treatment, but there are still conflicts in the literature. In humans there have been two therapeutic studies, one with NAC, and one with a combination of vitamins E and C, NAC, and selenium. Both studies had positive results.

Other forms of shock—due to blood loss and burns—are also characterized by severe oxidative stress. In animal experiments, antioxidants were effective treatments of shock due to blood loss and burns. In these studies new synthetic antioxidants such as Z-103 and Ebselen have been used. Preliminary tests in humans are under way.

ISCHEMIA/REPERFUSION   In any condition—such as stroke, heart attack, or during organ transplants—in which the blood flow to an organ is interrupted for a while and is then restarted, severe oxidative stress results. During a period of loss of normal blood flow, called ischemia, even for a period of a few minutes, the tissues become damaged by the lack of oxygen. But the main damage occurs during reperfusion, the restoration of the blood flow. A large release of reactive oxygen species caused by biochemical changes in the reperfused tissue results in severe oxidative stress and a lowering of intracellular glutathione levels. In a study of experimental ischemia of the heart in rats, antioxidants (SOD, CAT, vitamin E, and desferoxamine) offered effective protection, but the best was the synthetic agent H 290/51, which is 100 times as potent as vitamin E in inhibiting fat oxidation. In kidney transplants ischemia/reperfusion injury can be reversed by antioxidants. However, Lehr and Messmer have complained that, in spite of the strong scientific evidence that oxidative stress plays a key role in transplant surgery, "little attention is paid to the antioxidant status of patients undergoing organ transplants" [119]. This is an example of the slow pace at which the medical profession in general is actually putting these new insights to use.

In thrombolytic treatment for acute heart attacks in humans (in which the blood clot in the coronary artery is dissolved via a catheter inserted into the heart), patients develop a rise in fat oxidation and a fall in vitamin E and retinoids—signs of reperfusion oxidative stress. Thus, there may be a role for the use of some of the faster-acting antioxidants in thrombolytic treatment. Vitamin E acts too slowly to perform this task. A study of coronary artery bypass surgery showed that preoperative administration of vitamin E for four weeks resulted in fewer abnormalities in the electrocardiogram and fewer infarctions (heart attacks) around the time of the operation [189].

Patients with blocked leg arteries develop pain during exercise, called intermittent claudication, because of the lack of blood flow to support the exercise; they suffer from repeated ischemia/reperfusion damage every time they go for a walk. In these cases, after a period of exercise, there is a significant drop in antioxidant capacity in the blood. Such patients may need antioxidant supplementation.

There is also evidence that reactive oxygen species are a feature in cerebral vasospasm (contraction of the arteries of the brain). In animals antioxidants were effective in reducing vasospasm, particularly if applied locally. This might have some application in neurosurgery.

However, a warning note has yet again been struck. Paller and Eaton, in their study of reperfusion oxidative damage to the kidney, found, quite unexpectedly, that glutathione and one type of the antioxidant enzyme SOD were protective if given singly but highly toxic if given in combination [157], whereas glutathione and another type of SOD given together were synergistic. They advise that "great care must be used in designing and interpreting studies employing combinations of antioxidants." This applies particularly when these agents are administered intravenously.

CYSTIC FIBROSIS

In cystic fibrosis, the most common lethal inherited disease in the United States, a genetic mutation leads to abnormally sticky secretions by the mucous glands. The secretions gum up the lungs, resulting in chronic inflammation. Excessive activity of the white blood cells results in an increase in reactive oxygen species, which in turn promote excessive production of fibrous tissue in the lung. Malfunction of the pancreas results in an inability to absorb vitamin E properly from the intestines, which not only impairs the general antioxidant defenses of the body but also leads to liver damage. Moreover, the lungs of these patients get invaded by a bacterium (*pseudomonas aeruginosa*) that is a producer of reactive oxygen species.

Cystic fibrosis patients have very low beta-carotene and vitamin E levels and high levels of oxidized fats. Giving supplements of beta-carotene helps normalize the levels of fat oxidation. The lower the levels of the protective beta-carotene, the higher the levels of oxidation of the fats. Therefore, supplements containing beta-carotene could be of therapeutic value in cystic fibrosis. There is also a need to develop new synthetic antioxidants that can be absorbed in spite of pancreatic failure. Ramsey et al. say that "every patient with cystic fibrosis will eventually require supplementation with fat soluble vitamins" [169]. The potent antioxidant NAC can be given adequately by means of an aerosol spray. Winklhofer-Roob et al. have shown that patients with this disease who have low blood levels of vitamin C also have high blood indices of inflammation (MDA and TGFα levels), whereas those with high levels of vitamin C have "clearly lower values" of these indices of inflammation [232]. These researchers see these measurements as evidence that vitamin C is protective in cystic fibrosis, but they warn that, owing to the possible presence of free iron, further studies are needed to determine whether vitamin C supplements are safe. Free iron might be an important constituent of the fluid lining the respiratory tract in cystic fibrosis, in which case vitamin C, which becomes a potent oxidant in the presence of free iron, might be toxic.

Van der Vliet et al. offer another warning here [222]. It is theoretically possible that antioxidants benefit the bacterium more than the patient. The bacterium already produces one antioxidant itself (slimy alginate), which might interfere with the oxidant attack used by the white blood cells to kill the bacterium, as happens in acute eye infections. These investigators also warn about the possible danger presented by the interaction of vitamin C and free iron. Researchers developing an antioxidant therapy for cystic fibrosis must bear these caveats in mind.

THE
COMMON COLD

Twenty-five years ago Nobel Laureate Linus Pauling caused a furor by claiming that vitamin C could alleviate the symptoms of the common cold. Although the medical establishment greeted this claim with derision, many people today continue to reach for the vitamin C bottle as soon as signs of a cold start to develop. However, we now know that many of the symptoms of a cold are due not to the cold virus itself but to an overly enthusiastic response of the body's own immune system to the virus. During a cold the mucous membrane lining of the nose becomes filled with white blood cells, which secrete large amounts of their most powerful weapon, reactive oxygen species, aimed at killing the virus. Unfortunately, the reactive oxygen species produced by the macrophages of the immune system also attack the cells of the mucous membrane themselves, leading to a runny nose and discomfort. Thus, the strategy of giving antioxidants during a cold is not to kill the virus, but to weaken this attack by reactive oxygen species on the body's own tissues. However, vitamin C does also have some action in boosting immune responses [40, 103].

Hemilä and Herman have recently reexamined the findings of all twenty-one placebo-controlled trials carried out on this topic since 1970 in which 1 G or more of vitamin C a day was given [81]. They found that in every study there was no change in the number of colds; but there was a significant fall in the duration and severity of the symptoms. They criticize as inadequate some previous reviews that concluded that vitamin C did nothing for a cold. Although there is as yet no direct evidence to support this hypothesis, it is scientifically possible that vitamin C could reduce the symptoms of a cold while having no direct action on the cold virus itself.

ACUTE
INFECTIONS

Antioxidants are not, of course, cure-alls. Eye infections caused by bacteria are made worse by antioxidants, which interfere with the bactericidal action of the reactive oxygen species produced by the macrophages. This may well apply to other acute infections.

One certain exception is pneumococcal meningitis. In a study of rats, the antioxidant NAC protected against this infection [109] by reducing the brain edema and increased intracranial pressure produced by reactive oxygen species whose brain levels are increased by the infection and which are a major cause of death. The NAC did not interfere with the white cells' use of reactive oxygen species to kill the invading bacteria. In the case of nonbacterial eye inflammation, animal experiments show that antioxidants are protective. However, it would be wise to take antioxidants during the course of an acute bacterial infection only on the advice of a doctor.

BETA-THALASSEMIA   This disease causes severe anemia; consequently, patients must undergo repeated blood transfusions, which lead to severe iron overload and damage to many organs. The excess iron induces severe oxidative stress. Livrea et al. studied forty-two patients and found that blood indices of oxidation (conjugated dienes, MDA/TAB adducts, and protein carbonyls) were raised twofold [126]. Blood levels of many antioxidants were reduced (vitamin C by 44 percent, vitamin E by 42 percent, vitamin A by 44 percent, beta-carotene by 29 percent, and lycopene by 67 percent), presumably because they were used up by the iron-induced oxidative stress. The total antioxidant potential was reduced by 14 percent. The researchers suggested that this oxidative stress might be the cause of the myocardial damage that is the major cause of death in the disease. They concluded that administration of antioxidant compounds, but not vitamin C, could be beneficial. A previous attempt was made to treat this disease with vitamin E. Rachmilewitz et al. treated eight patients who had low vitamin E levels with 750–1,000 mg per day of vitamin E for sixteen months [161]. They found that there were some changes in antioxidant measures, but the treatment had no clinical effect. They concluded that vitamin E by itself might not be enough.

CIRRHOSIS
OF THE LIVER   Cirrhosis of the liver, which follows from

excessive intake of alcohol, may be due in part to fat oxidation and liver damage from reactive oxygen species. One trial of vitamin E supplements in established cirrhosis was ineffective, but this may have been another case of shutting the stable door after the horse had gone.

EFFECTS OF STRENUOUS EXERCISE

Moderate exercise, plus a low-fat, high-carbohydrate, and high-fiber diet, leads to a dramatic decrease in LDL (bad cholesterol) oxidation. However, it has been shown that extremely strenuous exercise leads to oxidative stress and compensatory rises in antioxidant enzymes and glutathione. This is because extreme exercise leads to muscle ischemia/reperfusion and a greatly increased use of oxygen. Athletes have higher levels of vitamin E in their red blood cells and vitamin C in their white blood cells than do nonathletes. Sen has written a good review of this topic [193]. He concluded that physical exercise is protective against oxidative stress in a number of ways. However, overdoing it can be harmful. According to Sen, people's physiological antioxidant status varies widely and thus a periodic assessment of one's susceptibility to oxidative stress would be desirable. After a review of the methods that have been used to combat oxidative stress during exercise and their results, he concluded that glutathione is ineffective probably because it does not get to where it is needed. NAC is converted into glutathione in the body and may be effective, but information is lacking on the effects of vitamins C and E. Only two tests have been carried out to see whether taking antioxidants actually increases athletic performance. The first, on swimmers, had a negative result. However, a trial of vitamin E in high-altitude mountaineering showed improved performance at low oxygen levels. At present we know only that athletes have better antioxidant defenses than do nonathletes. One wonders if athletes who take antioxidant supplements would come under the ban on artificial aids to improve athletic performance!

HYPERTHYROIDISM

Hyperthyroidism is associated with increased oxidative stress, leading to increased fat oxidation and lower serum vitamin E levels. Thyrotoxic muscle and heart lesions may be due in part due to oxidative stress. There have not as yet been any reports of the use of antioxidants in the disease.

INFLAMMATORY

BOWEL DISEASE   Lih-Brody et al. have studied biopsies from the intestines of patients with two such diseases—Chron's disease and ulcerative colitis [124]. Measuring indices of protein and DNA oxidation as well as levels of reactive oxygen species and the antioxidant enzyme SOD, they found evidence of severe oxidative stress. In Chron's disease they found raised levels of reactive oxygen species, iron, and SOD, as well as the presence of protein carbonyls (evidence of oxidative damage to proteins) and oxidatively damaged DNA. In ulcerative colitis there were raised levels of reactive oxygen species, protein carbonyls, and iron not only in the inflamed sites but in normal areas between the inflamed sites, showing that the raised levels were not just a result of inflammation. In ulcerative colitis there is free iron overload from the bleeding, so vitamin C must be used with caution.

MALARIA   Oxidative stress is involved in two ways in malaria. First, during an acute attack a great deal of pro-oxidant free iron is released from the red blood cells destroyed by the parasite; second, the defensive macrophages release many reactive oxygen species. Stimulating the formation of new blood cells rich in antioxidant enzymes is necessary to replace those lost during severe hemolysis in severe malaria.

On the other hand, the malarial parasite itself is very susceptible to oxidative stress. Thus, new pro-oxidant antimalarial drugs have been developed based on the traditional Chinese folk medicine qinghaosu. Furthermore, a diet high in fish oils rapidly produces a vitamin E deficiency, which results in slower growth of the malarial parasite; this, too, has been used in therapy.

MITOCHONDRIAL

DISEASES   Defects in parts of the cell called mitochondria, which provide the energy the cell needs to function, manifest as muscle and brain disturbances. An open trial of antioxidants (which included 2 G per day vitamin C and 400 mg per day of vitamin E) produced encouraging results [163]. The patients "appeared to" survive longer with less disability and fewer medical complications. Moreover, no side effects were reported. In another study nicotinamide (the anti-oxidant vitamin B_3) was given to good effect.

MYOTONIC DYSTROPHY

This disease is marked by raised plasma levels of reactive oxygen species and oxidized fats and lowered antioxidant defenses. No trials of antioxidant medication have been conducted as yet.

NEONATAL OXIDATIVE STRESS

Birth results in oxidative stress to the newborn infant, who passes rapidly from a low oxygen pressure in the uterus to a high one as soon as he or she starts to breathe air. A recent study of newborn rats compared the infants of mother rats given the antioxidant NAC with controls. In the control mother rats the oxidized glutathione levels in plasma increased elevenfold after birth; in the NAC-treated group the level increased only twofold. Evidently, much of the active form of glutathione was being depleted by the oxidative stress and converted into the oxidized form. This conversion was slowed down by the NAC. Human studies in this area are awaited with great interest because the birth process is fraught with danger to infants, many of whom suffer brain damage in which oxidative stress may play a part. Giving mothers anti-oxidants in the later stages of pregnancy may protect the infant from oxi-dative stress at birth.

PANCREATITIS

Inflammation of the pancreas is often a result of digestion of the gland by its own highly potent digestive juices. There are 20,000 cases of this syndrome in the United States every year. In sixteen out of twenty-four animal experiments, antioxidants alleviated acute experimental pancreatitis. In humans, five studies have shown that this disease is associated with severe oxidative stress and depletion of the antioxidant defenses beta-carotene and vitamins C and E. One clin-ical trial of NAC has produced promising results.

PREECLAMPSIA   This complication of pregnancy, the leading cause of maternal mortality in the West, is marked by high blood pressure and edema (swelling). In both plasma and cerebrospinal fluid, the total number of antioxidants is above the normal levels seen in uncomplicated labor. This is mainly because of the raised levels of the plasma antioxidant uric acid, together with some as yet unidentified agents. Weight for weight, the predominant antioxidant in blood is uric acid, whereas in the cerebrospinal fluid it is vitamin C. There is also increased consumption of vitamin C by the body but not vitamin E [93]. Serum levels of the antioxidant enzyme GSHpx are decreased, together with signs of increased lipid oxidation [164]. Oxidative stress may play a role in preeclampsia, but as just one of a number of factors involved.

RENAL

DIALYSIS   The red blood cells in patients under-going dialysis suffer oxidative stress, with reduced levels of the antioxidant enzymes SOD, CAT, and GSHpx. This results in the red blood cells' dying before they should, because they have more rigid (and thus brittle) cell membranes and are subject to increased fat oxidation. Antioxidants added to the perfusion fluid would be expected to be advantageous. Another kidney ailment, glomerular nephritis, is associated with oxida-tive stress [219]. Undialyzed patients with chronic renal failure show a failure in their antioxidant defenses that worsens as the renal failure increases with time [29]; subsequent dialysis worsens the condition, which may account for the high rate of atherosclerosis in these patients.

RHEUMATOID

ARTHRITIS   This is an autoimmune disease in which the immune system mistakes the joint linings for foreign invaders and pro-ceeds to launch an attack upon them by reactive oxygen species. The fluid in the affected joints in these patients has elevated levels of reactive oxy-gen species and free iron. The more severe the disease the higher are mea-sures of fat oxidation in the blood. Many patients with rheumatoid arthritis are marginally deficient in vitamins C and E. A low blood level of vitamin E, beta-carotene, and selenium is associated with an eight-fold increase in the risk for the disease [11]. Panetta et al. of the Lilly Research Laboratories are carrying out a study of new synthetic drugs, such as 4-thiazolidinone, which have potent antioxidant and anti-inflammatory actions in this disease [159].

SYSTEMIC SCLEROSIS   This is a connective tissue disease (also called scleroderma) due to overproduction of collagen that leads to vascular damage and Raynaud's syndrome (repeated attacks of vascular spasm leading to gangrene of the extremities). It is marked by repeated ischemia and reperfusion in the tissues with resulting oxidative tissue damage. Many organs are involved, including the heart, lung, intestines, and kidneys. These patients have low plasma levels of vitamin C but normal levels of vitamin E. It is not clear if the low plasma levels of vitamin C—due to neither dietary deficiencies nor malabsorption—is a cause or result of the oxidative stress. Antioxidants might well be effective in this disease.

TUBERCULOSIS   The interesting fact has been unearthed that the bacterium responsible for the disease has a defect in one of the genes that regulates its antioxidant defenses. Thus, in tuberculosis, these defenses are low. This explains why the oxidant drug isoniazid is so effective in treating the illness. Related organisms that do not have this defective gene are not sensitive to isoniazid.

clinical data on antioxidants

FLAVONOIDS

SOY.   In seventeen out of twenty-five animal studies, genistin, a component of soy, showed anticancer activity. In China and Japan, where soy consumption is high, there are low rates of cancer of the breast, prostate, and colon. However, epidemiological studies looking at the correlation between high soy intake and cancer rate have been inconsistent. Out of twenty-six studies, ten reported positive results, fifteen no effect, and one a negative effect. So at least the balance is in the positive direction.

RED WINE. Goldberg expresses the opinion that, if we all drank two glasses of red wine a day, the incidence of coronary heart disease would fall by 40 percent [69, 70]. However, drinking wine in excess may lead to cirrhosis of the liver, and "the use of alcohol for cardiovascular purposes should not be encouraged as a public health measure" [36]. Moreover, alcohol itself is a pro-oxidant, and LDL (bad cholesterol) is more easily oxidized in beer drinkers.

A skeptical note was introduced by a report from a country with high beer consumption (Holland). De Rijka et al. gave twenty-four healthy people with normal blood fats red wine supplements for four weeks [42]. The level of plasma antioxidants and fats was unchanged. However, this trial was probably too short. A contrary report states that consumption of red wine does reduce the susceptibility of LDLs to oxidation [59].

TEA. In one study a mixture of antioxidants from green tea had a potent inhibitory effect on the growth of mammary cancers in rats. A study of over 100,000 people in Holland showed that consumption of black tea for more than four years did not lower the incidence of stomach, colorectal, or breast cancer. However, black tea (but not green tea) also contains tannins, which can promote tumors. A prospective study of 35,369 women in the Iowa Health Study found that the intake of two cups of "nonherbal" tea a day (equivalent, in amount of antioxidants therein, to one helping of fruit or vegetables) was related to a significant 40 to 70 percent reduction in the incidence rate of two types of cancer only—digestive tract and urinary tract [240]. The authors of the study quote a number of other studies showing protection by tea against oral, pharyngeal, and nasopharyngeal cancers. They suggest that the flavonoids in tea may block the cancer-producing nitrosamines in the intestines. They also mention that the kidney might concentrate the flavonoids, which will result in higher urinary levels of flavonoids and thereby help in fighting tumors in the renal tract. They warn, however, that drinking tea that is too hot might increase the risk of esophageal cancer. The protective effect of the large tea consumption in Japan may be responsible for lower lung cancer rates in the Japanese, despite their higher rate of smoking than Americans'. An epidemiological study in Japan reported that cancer patients who drank more than ten cups of green tea a day had an increased life expectancy (by 4.5 years for men and 6.5 years for women) as compared with cancer patients who drank fewer than three cups a day [60].

In the case of coronary heart disease a study in Wales had a negative result [87]. The investigators selected 1,900 males in Caerphilly and followed them for fourteen years. There was no correlation between the amount of black tea drunk and the incidence of heart attacks. In fact, those people who drank more tea had an increased general mortality rate. Further research showed that tea drinking in Wales is associated with a less healthy lifestyle (obesity, smoking), which was the probable cause of the increased death rate—an example of a confounding variable at work. However, the Welsh study did find a protective effect of onions, which also contain protective flavonoids.

GARLIC. Laboratory tests have shown that garlic extracts protect LDLs against fat oxidation and have antiviral, antimicrobic, and anticancer properties. Experiments in animals show that garlic extracts slow the development of atherosclerosis [50]. Garlic also protects against irregularities in the heartbeat and prevents the development of fatty streaks on coronary vessels (a prelude to atherosclerosis) in rabbits fed large amounts of cholesterol.

Five epidemiological studies (in China, Italy, Poland, Australia, and the United States) have shown that a diet high in allium vegetables (i.e., garlic, onion, chives, shallots, etc.) that contain allicin and related antioxidants protected against stomach and colon cancer but not against breast and lung cancer.

In an experimental study of ischemic brain damage in rats, garlic extract reduced the size of the resulting brain damage and reduced brain reactive oxygen species production in the treated rats as compared with controls.

The bacterium (Helicobacter pylori) responsible for gastric and duodenal (peptic) ulcers is very sensitive to the antibacterial effect of garlic

[200]. The authors suggest that cases of peptic ulcer that are resistant to the antibiotics now used to treat the disease might benefit from the equivalent of two small cloves of garlic a day.

OLIVE OIL. Trichopoulou suggests that olive oil may offer a "modest protection" against breast, ovarian, and lung cancer, mainly because, in addition to antioxidants, it contains monounsaturated fats, which are more protective than either saturated animal fats or polyunsaturated fats found in many vegetable sources [218].

the safety of antioxidants

and recommendations

The first rule for physicians Hippocrates laid down was "Thou shalt do no harm." Therefore, it is of prime importance that we find out if antioxidants are harmless before we prescribe them to our patients. The first section of part 3 examines this question in depth. The second section discusses what kind of diet is indicated by the data, whether people should or should not take supplements of antioxidants in addition to what they get in the diet, and, if they do take them, under what circumstances they should do so.

safety of antioxidants

Just how safe are the common antioxidants in current use? Opinions vary wildly from the belief that they are completely safe to protestations that under certain conditions they can be highly dangerous and even lethal. What follows are the actual data on the safety of antioxidants and a discussion of the opinions offered by various experts.

part 3

BETA-CAROTENE   Several authorities have judged beta-carotene to be safe or minimally hazardous. However, it is clear from the results of the CARET and ATBC trials reviewed above that beta-carotene supplements given by themselves should not be used in smokers, in whom the supplements have been reported to raise the death rate from lung cancer, heart attacks, and other conditions. Nor should they be used in people who have had a heart attack. Animal studies suggest that beta-carotene can also increase liver damage caused by alcohol. It should therefore be given with caution to people who drink a lot of alcohol. Furthermore, raising beta-carotene intake lowers the serum levels of the important antioxidant carotenoids, lutein, and zeaxanthin. This is especially relevant in the eye disease macular degeneration (see the section on eye diseases), because beta-carotene is not found in the retina, whereas lutein and zeaxanthin are essential for proper retinal function. Giving beta-carotene in this condition would tend to deprive the retina of getting the lutein and zeaxanthin it needs; the latter two should be added to any vitamin supplement formula. Another point to note is that canthaxanthin supplements, taken in large amounts for a long time, can cause crystals to form in the retina that interfere with vision. Fortunately, the crystals are absorbed upon stopping the supplement.

The yellowing of the skin caused by beta-carotene is benign and completely reversible, and in any case it looks more like a good suntan than jaundice.

VITAMIN C   Because Pauling made megadoses of vitamin C notorious, there has been greater concern about the possible toxic side effects of vitamin C taken in large doses than of the other antioxidants. In most people doses up to 500 mg per day seem to be perfectly safe, yet there may be some exceptions.

One problem is oxalate kidney stones. The chief metabolite of vitamin C is oxalic acid, the main ingredient in the commonest type of kidney stone in the West (80 percent). Chalmers et al. gave seventeen kidney stone patients and eleven normal controls 2 G per day of vitamin C by mouth, and reported that the former group excreted more oxalate and less vitamin C in their urine than did the controls [30]. When the vitamin C was given intravenously, this did not occur. The authors suggest that the stone formers are not able to absorb vitamin C as well as normals can; vitamin C taken orally thus remains in the intestines longer, forming more oxalate, which is absorbed as such. They advise that people who form oxalate stones should not take vitamin C supplements. Urivetsky et al., whose study of fifteen kidney stone patients given 2 G per day of vitamin C revealed a twofold increase in the excretion of oxalate in the urine, concluded that the chronic administration of vitamin C supplements made them more liable to form stones [220]. They advise that such people should not take more than 500 mg per day of vitamin C. Rivers agrees that vitamin C poses no problem to normal people, but that oxalate stone formers should not take vitamin C supplements [177]. Goldfarb does not agree with this, contending that normally the metabolic pathway from vitamin C to oxalate is working at full blast anyway, so that any increase in vitamin C intake does not lead to any increase in oxalate excretion in the urine [77]. The excess vitamin C is simply excreted as vitamin C in the urine. However, in a few susceptible individuals a marked increase in urinary oxalate can follow the ingestion of 1–2 G of vitamin C. Schmidt et al. gave four patients a very large dose (10 G per day) of vitamin C and found that the mean oxalate excretion increased from 50 mg per day to only 87 mg per day [189]. They concluded that this was equivalent to the magnitude of effect that a simple change of diet could produce. Diplock states that the "stone story" has proven on critical examination to be "without foundation" [44]. If excess vitamin C is ingested, it is excreted as such, not as oxalate. He attributes earlier studies' apparent findings that increasing vitamin C intake led to increased oxalate output to technical errors in the estimation, such as allowing the urine to become too alkaline. In fact, he concludes that vitamin C in the doses normally used in supplements is entirely free from side effects. Certainly I have not seen any reports of a case in which a kidney stone, in real life as opposed to theory, could be traced to vitamin C supplements. Nevertheless, high doses of vitamin C should be given to known oxalate stone formers with caution and only at the discretion of the individual physician.

Another possible problem with vitamin C is that, in the presence of free iron, it becomes a pro-oxidant and should therefore not be given to people who may have free iron that might react with the vitamin C.

Normally, iron in the body is safely stored inside various proteins, and free iron levels are extremely low. This is just as well, as free iron is exceedingly toxic. It thus causes no problems in this regard in normal people. But some experts (particularly Herbert [83–85]) advise that vitamin C supplements not be taken by patients with iron overload, as in hemachromatosis, sideroblastic anemia, and thalassemia. Herbert suggests that vitamin C supplements should not be taken if the blood level of ferritin, an iron-binding protein, is over 120 μg/L. In contrast, other authorities state that iron is always safely bound to protein; so vitamin C in fact, as opposed to theory, never interacts with free iron to produce pro-oxidant effects in real life [14, 38, 45, 147]. Herbert refers to a submission paper made by M. Krikker to an FDA committee that mentions "several deaths in athletes due to this cause." I am not aware of any such interaction reported in any scientific paper. He also states that patients in a study on the effects of vitamin C on cancer by Moertel et al. [145] at the Mayo Clinic did "much worse" on vitamin C. I have read the latter paper and was unable to interpret it in this way. In fact, the Mayo Clinic group treated 100 terminally ill cancer patients with 10 G per day of vitamin C for up to eighteen months (average four months) but the treatment had no effect of any sort, good or bad. The study was carried out to refute the claim made by Cameron and Pauling that large doses of vitamin C produced marked benefit in such cases. Moertel et al. came to the conclusion that this claim was based on a simple error in the design of the trial [145]. Daily and Zemel, in an editorial in the *American Journal of Clinical Nutrition*, characterize Herbert's contribution to the debate as "polarizing hyperbole and grandstanding. . . . What is certain is that the continuing debate is not well served by being reduced to the type of diatribe evident in Herbert's commentary" [37]. In a paper read to a meeting of the Oxygen Club of California, Gladys Block presented evidence to show that vitamin C, even in the presence of free iron in living humans, actually inhibits fat oxidation.

However, Barton and Bertoli point out (as Herbert has claimed over the years) that hemachromatosis is a much more common disease than is generally recognized [10]. One million Americans are homozygous for the gene responsible and one in eight of the population are heterozygous. Iron overload problems are commonly clinically overlooked and screening tests [49] should be much more widely used [10]. Such tests would therefore seem to be advisable before a doctor's recommending high doses of vitamin C.

There are conflicting reports as to whether vitamin C supplements may induce or improve cataracts in elderly patients with diabetes. In most people, as we saw earlier, vitamin C appears to protect against cataracts and against diabetes. Elderly diabetics may prove an exception to this rule.

To avoid hemolysis, vitamin C should not be given intravenously to patients with a condition known as glucose-6-phosphate dehydrogenase deficiency [122]. Bendich and Langseth state that some previous anecdotes claiming the alleged toxicity of vitamin C have been exaggerated [12]. In controlled trials it has been shown not to destroy vitamin B_{12}, it is not mutagenic, it does not cause rebound scurvy or abnormal psychological reactions, and it does not impair copper utilization—all of which had been claimed previously on an anecdotal basis. However, it can interfere with a number of clinical and laboratory tests [143], including blood level tests. The tests for glucose, uric acid, creatinine, alkaline phosphatase, and inorganic phosophate give falsely high levels. The tests for bilirubin, lactate dehydrogenase, carbon dioxide, potassium, catechol-o-methyl transferase, monoamine oxidase, cholesterol, creatine kinase, and dopa-β-hydroxylase give falsely low levels. The test for occult blood in feces gives a false negative result. Tests for acetaminophen gives false positive results. A number of trials of large doses of vitamin C (for example, those by Paolisso et al. [160], who gave 1 G per day for four months to forty-nine people; Bussey et al. [25], who gave 3 G per day for two years to forty-nine people; and McKeowen-Eyssen et al. [136], who gave 400 mg per day for two years to ninety-six people) have reported no significant side effects.

In animal experiments vitamin C can have either anticancer or cancer-promoting effects, depending on circumstances. Vitamin C depresses DNA, RNA, and protein synthesis (i.e., cell growth and division) in some cancer (neuroblastoma) cells that have high iron levels, but it does not do so in normal cells. The high iron levels cause the vitamin C to have a pro-oxidant effect. This suggests that vitamin C might be a powerful enhancer of some antitumor drugs used in the treatment of neuroblastoma. Vitamin C causes DNA damage in neuroblastoma cells but not normal cells. Benzathione (a powerful carcinogen) toxicity in guinea pigs is reduced by vitamin C.

In one study, whereas an antioxidant mixture of beta-carotene, vitamins C and E, and glutathione was effective in preventing the production of oral cancers by cancer-producing chemicals in hamsters, vitamin C given by itself actually increased the cancer rate [194]. The researchers suggested this result might be due to the production of toxic derivatives of vitamin C that are normally mopped up by vitamin E and the other antioxidants in the mixture. Without this mopping-up operation, vitamin C by itself proved to be harmful. This result stresses once again the enormous importance of the team action of the antioxidants. This is further stressed by Prasad and Kumar, who state that multiple antioxidant administration is essential for the maximum reduction of cancer incidence in a high-risk population and that the use of just one or two antioxidants is likely to be ineffective and even harmful [165].

However, a study by Jacques et al. suggests that this is not an absolute rule [96]. They measured the effects on blood levels of the raised intake of only one of either vitamin C, vitamin E, or carotenoids; giving only one led to raised blood levels of one or both of the other two without any impairment of antioxidant status. They explained this result by postulating various interactions between the different antioxidants at several levels—during the preparation of food, in the gastrointestinal tract, in the cell, and by sparing. For example, the raised vitamin C level may convert, by its synergistic action, more inactivated oxidized vitamin E to the active form of vitamin E.

VITAMIN E   There are only two recognized complications of vitamin E therapy. The first is that it can worsen the coagulation defect caused by vitamin K deficiency, which is due either to malabsorption of vitamin K from the intestines or to anticoagulant therapy. The second is that, whereas vitamin E certainly lowers blood platelet counts, this may not always be beneficial in certain people, as it may increase the risk of hemorrhagic stroke. Therefore, so long as these are recognized, vitamin E therapy is considered safe by most authorities [53, 138, 187, 226]. Byers and Bowman, however, worry about possible as yet unknown side effects of long- term, high-dose vitamin E therapy [26]. This worry does not seem to be justified in view of the fact that many hundreds of thousands of people have taken vitamin E supplements over the last decade without any reports of such complications. However, raising the level of intake of alpha-tocopherol (vitamin E) lowers the level of its close relative gamma-tocopherol. Gamma-tocopherol is more potent than alpha-tocopherol in detoxifying nitrogen dioxide (the potent oxidant in smog and cigarette smoke); thus, gamma-tocopherol should be an ingredient in any antioxidant program.

recommendations   What conclusions can we draw from this mass of data? Given that many of the investigations and experiments have had contradictory and confusing results, clearly much more work needs to be done to clear up some of the issues. However, certain things have already been clearly established.

It can now be taken as a fact that toxic reactive oxygen species play a significant role in many acute and chronic diseases such as inflammation, coronary heart disease, cancer, diabetes, cystic fibrosis, rheumatoid arthritis, Alzheimer's disease, Parkinson's disease, and many more. Furthermore, they play an important role in many normal bodily processes, such as control of gene expression, immune responses, control of brain synapses, the action of white blood cells in killing bacteria, and others. We now know quite a lot about the nature of the body's defenses against reactive oxygen species. More than twenty important antioxidant systems in the body have been identified.

It is also now beyond dispute that a healthy diet should be designed to have a plentiful and varied supply of fruits and vegetables, as well as a proper attention to the right and wrong sorts of fat, reduced salt, and a low alcohol and low calorific intake. However, even today, it has been estimated that only 10 percent of the U.S. population follow this advice with regard to fruits and vegetables. Dietary habits change slowly. In many areas fruits and vegetables are expensive, and some people cannot afford them. Many restaurants in the United States are dietetic disaster areas, with red meat, salty potatoes, and buttery breads as the staple foods. In San Diego, California, the Yellow Pages list only eight vegetarian restaurants.

In home kitchens many vegetables tend to be overcooked, which destroys their vitamin C. How much easier to buy a bottle of tablets labeled "antioxidant vitamins" and trust that will do the trick!

Furthermore, the general public is currently much confused by all the contradictory reports that have been presented by the media. One day beta-carotene is in, the next it is out. One day margarine is recommended instead of butter; the next day a different advice is given following tentative research reports by scientists. Many books written for the general public on this topic are uncritical and misleading.

As it is now clear that antioxidants work as a team, it is pointless to design large and expensive clinical trials of just one or two of the antioxidants. The antioxidant mixture to be tested should contain all the prominent antioxidants, both water-soluble and fat-soluble, various carotenes and flavonoids, and possibly agents designed to raise glutathione levels in the body.

Most excess minerals added to the supplement are simply not absorbed, except in cases of mineral deficiency from the previous diet. Selenium is an exception, but selenium overdosage is toxic. However, calcium and magnesium are beneficent even when not absorbed, as they form very insoluble salts with deleterious fatty acids in the intestines and so prevent their absorption from the intestines. This may help to prevent colorectal cancer. Calcium is also required by some people to guard against osteoporosis.

Furthermore, there is now a very strong case that antioxidants are needed in the medical treatment of specific conditions such as ischemia/reperfusion, acute myocardial infarction, asthma, shock, ARDS, cystic fibrosis, diabetes, Alzheimer's disease, and so on. Further research may add diseases like Parkinson's, schizophrenia, ALS, and AIDS to this list, as the evidence already looks promising.

The present major disagreement centers on whether antioxidant supplements should be recommended for the general population in an attempt to ward off the development of chronic diseases such as heart disease, cancer, cataract, and Parkinson's disease. In other words, should we raise the RDAs for the key antioxidant vitamins?

The following is an account of the most up-to-date recommendations made by recognized experts in the field, beginning with those who advise against raising the present RDAs.

Meyers and Maloley require "clear proof" of efficacy and the lack of long-term toxicity before they can recommend supplements of vitamins C and E [143]. They maintain that many of the research reports that suggest that such supplements are needed are flawed by not taking into account other risk factors for heart disease and other complicating health-conscious behaviors (healthy lifestyles) by the subjects. However, as we have seen, many positive trials cannot be explained on the healthy-lifestyle hypothesis. Meyers and Maloley estimate that, in spite of their advice, some 25 percent of Americans self-medicate with vitamin supplements on a regular basis.

Illingworth agrees with Meyers and Maloley and offers this caveat: "In the absence of more convincing clinical trial data it seems premature to advocate supplementation with antioxidant vitamins [to prevent coronary heart disease] except in selected high risk patients" [94].

Oliver says "there is a sound scientific basis and rationale for increasing the intakes of vitamin E and C to reduce oxidation of LDL" to prevent coronary artery disease, but he concludes that it is still too early to recommend supplements. "The best possible advice" is to eat foods high in vitamins C and/or E, including cereal oils, nuts, fresh citrus fruits, fresh vegetables, and potatoes—many of which are expensive [151]. Potatoes, which are affordable, are lacking in antioxidants other than vitamin C, and food sources rich in E are all fatty.

Rautalahti and Huttunen of Finland's National Public Health Institute, Helsinki, advise against antioxidant supplements for the prevention of cancer because of the lack of agreement in the clinical research base [171]. However, the trials they mention are hardly a fair selection of the database. They select two negative trials, one in melanoma and the other the flawed ATBC trial in Finland, where we have seen that inadequate doses of synthetic vitamin E were used in the hopeless task of trying to roll back the effects of heavy cigarette smoking over half a lifetime. The positive trial they cite is the Linxian trial in China, in which the subjects were nutritionally deprived anyway [18]. They should have used a better and wider database.

Maxwell [132] and Hoffman and Garewal [90], in reviews of atherosclerosis, say that though research to date has yielded promising and exciting results they cannot advocate antioxidant supplements—again largely on the basis of the flawed ATBC trial.

Diplock says we need to look at three levels of evidence [143]:

1. epidemiological studies of the relation between levels of intake of nutrients and disease rates,
2. prospective studies comparing plasma levels of nutrients and disease rates,
3. double-blind, placebo-controlled intervention trials.

He concludes that there is ample evidence from levels 1 and 2 to link antioxidants and coronary heart disease and some forms of cancer. But there is as yet little evidence from level 3.

The disagreement among doctors on this matter is well illustrated in the pages of the twentieth edition of the prestigious *Cecil's Text Book of Medicine*, published in 1996. In his chapter entitled "Cancer Prevention," Gilbert Omenn hardly mentions antioxidants. In his discussion of diet he mentions only the role of fat and fiber and ignores the extensive literature I have quoted on the role of antioxidants. In his discussion of tests of supplements of antioxidant vitamins in cancer, he mentions only the Finnish ATBC trial, but not its many defects. In contrast, William Blot, in his chapter on "The Epidemiology of Cancer," gives prominence to the role of fruits and vegetables and says that carotenoids as well as vitamins C and E can reduce the risk of getting cancer. Joel Mason, in his chapter on "Consequences of Altered Micronutrient States," also states that beta-carotene and vitamin E are protective against some cancers. Roland Weinsier, in his chapter on diet, states that "unprescribed daily use of supplements in amounts exceeding the recommended daily allowance should be avoided." I agree here with the word "unprescribed." The complexities of antioxidant therapy are such that it should not be used without medical advice, as I explain later in this section.

Arguments for the other position—that RDAs should be changed now and that antioxidant supplements have a place in treatment—have come from some equally eminent authorities.

In an editorial in the *Journal of the American College of Nutrition*, Blumberg expresses the contrary position forcefully [19]: "It is unrealistic and unnecessary to wait until the clinical trials are complete before applications for disease prevention are endorsed." He says that the evidence should be judged in toto and not just on the standards applied to new drugs—that is, prospective, randomized, double-blind, placebo-controlled trials. He also comments that antioxidants "appear remarkably benign even at high supplementary intakes" and that they are inexpensive to boot. He concludes: "Recommendation to wait until every conceivable study has been designed and conducted to achieve a level of absolute certainty will result in the continuing cost of the disease to the individual and society."

Gey agrees that an updated prudent diet should contain more than the current RDAs of vitamins C and E and beta-carotene plus supporting plant antioxidants [64]. The intake of vitamin C needed to prevent scurvy is 1 mg per kg of body weight per day, whereas most other animals make vitamin C at a much higher rate (40–275 mg/kg/day, or 3–19 G/day for a human weighing 70 kg!). Moreover, it has been estimated that our Stone Age ancestors, who lived by hunting and gathering, had a vitamin C intake of about 325 mg per day. It is therefore likely that evolution has produced a human body that needs this amount of vitamin C.

Mehra et al. state, "It should be recognized that 'definitive data' is often not available to support all decisions in medical practice" [138]. They routinely recommend antioxidant supplements (beta-carotene and vitamins C and E) for the primary and secondary prevention of atherosclerosis. But they stress that this is not a substitute for a rigorous program of reduction of all the other known risk factors. Leske et al. give similar advice for cancer prevention [121]. Weisburger of the American Health Foundation says the concept of RDAs, with its focus on the avoidance of deficiency diseases, is out of date [231]. He says that it should be replaced by the new concept of optimal nutrition to avoid chronic diseases and to protect against environmental toxins. Byers and Bowman say "better dietary advice, possible fortification of food supplies, and the use of rationally formulated nutritional supplements may soon emerge as public health strategies to help prevent chronic diseases" [26].

However, optimal levels of intake have yet to be defined. On this point Tengerdy says, in the case of vitamin E, that the optimum dose depends on many factors and needs to be determined for each person [215]. In particular, because of the demonstrated effect of vitamin E in boosting immune responses, he recommends vitamin E supplementation for three to four weeks before vaccination, provided the blood level of the vitamin is monitored and an adequate selenium intake is assured. An "updated prudent diet," according to Gey, should contain 1 mg of vitamin A, 6–15 mg of beta-carotene, 60–250 mg of vitamin C, and 60–100 mg of vitamin E supported by "plant antioxidants" [64]. Toxic levels are estimated by Van der Hagen et al. as over 150,000 IU for beta-carotene, 1 G for vitamin C, 800 mg for vitamin E, and 500 µg for selenium [221]. These authors support the use of supplements for the prevention of eye disease. Barber and Harris advise the following levels of intake of antioxidants: beta-carotene 15–30 mg; vitamin C 100–500 mg; vitamin E 200–800 mg; selenium 10–100 µg [9]. Hathcock says that to inhibit the formation of cancer-producing nitrosamines in the stomach completely, one needs 1 G per day of vitamin C [80]. He also notes that very high levels of ingestion of vitamin C can suppress the absorption of copper and that high levels of zinc ingestion (100–300 mg per day) can raise LDL and lower HDL levels.

In response to those who recommend relying entirely on getting people to change their diet, Johnson says that experienced clinicians know that people rarely change their dietary habits; therefore, this approach is "likely to be the least acceptable, or possible, or successful for many high-risk individuals" [98]. This sentiment is supported by Weisburger, who recommends vitamin supplementation for people who do not, will not, or cannot afford to eat their three servings of vegetables and two of fruit a day [230].

Levine et al. point out five flaws in the current method for determining the RDA of vitamin C and list eight criteria as to how it should properly be done [122]. They state that the current RDAs do not reflect the optimum dosage, but we do not yet know enough to say exactly what the optimum amount should be; they suggest provisionally between 200–500 mg per day of vitamin C, as much as possible from fruits and vegetables. As we saw earlier, the diet of our Paleolithic ancestors has been estimated to contain around 325 mg of vitamin C per day. The maximum amount obtainable from a normal diet is around 500 mg per day.

An important paper by Levine et al. provides the strongest evidence to date that the RDA for vitamin C should be raised [123]. They studied seven healthy volunteers over six months on various doses of vitamin C—

from 30 to 2,500 mg per day—and measured how much vitamin C in the diet is needed to obtain a maximum blood level. They found that the present RDA of 60 mg per day does not do this; 100 mg per day are required to saturate the blood cells (red and white), which show a concentration fourteen times that of plasma levels. But 200 mg per day are needed to achieve satisfactory plasma levels. So they recommend that the RDA for healthy young men be raised to 200 mg per day. They state that any intake over 500 mg per day is simply excreted in the urine and therefore wasted, but that levels up to 1 G per day are safe; above that level oxalate and urate levels in the urine increase. However, they stress that these recommendations apply only to healthy young men; different levels may be needed for women, smokers, the sick, and the elderly.

Block has published a cogent criticism of the orthodox view that only double-blind, placebo-controlled trials can supply the answer to this question [16]. Listing a series of fallacies in this approach, she says these trials usually:

- select persons at high risk for a disease;
- rarely test more than one or two substances and usually at a single dose;
- test only the efficacy of an agent given for a limited time, usually late in life;
- tell us little about prevention of long-term chronic diseases;
- tell us nothing about whether the agent at high dose might reduce the risk of chronic diseases if taken throughout a lifetime;
- tell us nothing about the combination of antioxidants, which we have seen to be so important;
- do nothing to resolve the questions that interest us, which involve persons with no unusual risk of disease, a lifetime exposure to noxious and protective agents involving an enormously complex interaction among nutrients, and the effects of these nutrients on hundreds of diseases, many uncommon.

Clinical trials, she says, simply cannot answer these questions. What we need is a solid examination of the laboratory and epidemiological evidence.

My additional comments are that many present-day recommendations apply only to people not under oxidative stress, that they do not address the question of teamwork between antioxidants, and that they do not con-sider the possible prevention of a number of chronic diseases as opposed to only one. Furthermore, if the intake of vitamin E is raised to 400–800 mg per day, then more vitamins C and B$_3$ (nicotinamide) will be needed as a vitamin E helpers.

Draper and Bettger say that the question of whether RDAs should be raised is a pharmacological matter and not one for nutritionists to dis-cuss [45]. Floren et al. place the responsibility on the individual physi-cian, who should rely on his or her clinical judgment as to whether to prescribe vitamin supplements [56]. The trouble is that many clinicians do not have access to reliable information on which to base their clini-cal judgment—that gap being one of this book's intentions to fill.

A 1997 survey based in New Orleans was made by a questionnaire that asked cardiologists if they took antioxidant supplements [139]. Forty-four percent of those who responded to the questionnaire said that they did (in the daily range of 400 mg of vitamin E, 500 mg of vitamin C, and 20 mg of beta-carotene). Ironically, only 37 percent of them said that they recommended these supplements for their patients.

A new approach is now being explored: how to enrich our food with antioxidant and protective agents. The simpler approach is to add anti-oxidant vitamins to basic foods. Studies address the genetic engineering of food to prevent heart disease and cancer [105], and such measures as growing garlic with selenium fertilization [95]. The authors of these stud-ies state that "in view of the impossible task of persuading the public to eat only those foods that are presumably good for their health . . . the time has come to enrich our foods with known cancer preventive agents so that their benefit can be realized fully over the life span of the indi-vidual" [95]. Their gloomy estimation of the potential to persuade the public to improve its diet is supported by a study of an intensive, two-year educational and exhortation campaign in which investigators operated from a selected supermarket. They had only "modest" results in getting the subjects of the experiment to buy more healthy foods. The main improvement lay in increasing the purchase of fruits and vegetables. Measurement of any further changes in nutritional knowledge and atti-tudes showed only "most modest" improvements [179].

conclusions In spite of the many disagreements in the literature over details, the total evidence from all sources seems to indicate that antioxidants can play a role in the prevention and treatment of many diseases. Most studies report positive effects in some areas even if they do not agree on the details. There are very few reports of actual harmful results of giving antioxidants—in fact as opposed to theory. In those that do, the cause of the harmful result is usually clear and can be avoided.

However, the main problem remains, whether we should rely on diet, or supplements, or a combination thereof. Obviously, in the best of all possible worlds the best solution would be diet, with the proviso that vitamin E supplements should be taken so that a person can avoid consuming too much fat from vitamin E–rich foods. This in turn requires taking enough vitamin E helpers, such as vitamins B_3 and C and selenium. I do not see how anyone who has studied the data on vitamin E in the prevention of heart disease (given in part 2) could fail to recommend that the RDA for this antioxidant vitamin be raised to at least 400 mg per day. The well-conducted investigations with positive results showing the protective role of vitamin E cannot simply be ignored. A good review of this topic has been given by Weber et al. [228]. Furthermore, a diet high in fruits and vegetables does more than provide antioxidants. As Wiseman and Halliwell say in their review, "it would be naive in the extreme to assume that the protective effects of fruits and vegetables are related only to their antioxidant content" [234]. Other important factors include antiangiogenesis factors, inducers of enzymes that combat carcinogens, fiber and phytates, as well as a reduction of the intake of fat and iron. On the other hand, some antioxidants may in fact exert some of their benefits by means other than their antioxidant effect—for example, the likely action of vitamin C in mopping up carcinogenic nitrosamines in the stomach.

Furthermore, the simple admonishment to eat more fruits and vegetables is not enough—which fruits and vegetables? Again, prudent advice would be a wide variety, including sources of carotenes (carrots, apricots, spinach, red and yellow peppers, and collard greens for beta-carotene; dark green leafy vegetables for lutein and zeaxanthin; tomatoes and peanuts for lycopene); of flavonoids and polyphenols (onions, green tea, olive oil, red wine, peanuts, coffee, chocolate, oranges, licorice, and many others); of vitamin E (nuts, grains, margarine, mayonnaise, whole wheat germ, dark green leafy vegetables); and vitamin C (potatoes, citrus fruits, and others), together with olive oil whenever possible, garlic and shallots (for allicins), and the spices and herbs listed earlier that contain a rich variety of antioxidants (see the appendix for more details). Note that potatoes do not contain significant amounts of antioxidants other than vitamin C.

Health authorities need to mount a much more vigorous campaign to inform the public of the enormous importance of a massive nationwide change in eating habits, in particular with regard to fruits and vegetables. It is somewhat depressing that Patterson et al. made precisely this same point in 1990 [162]. Two years later Blackburn pointed out that the fruit and vegetable industry had showed lack of leadership by failing to promote their wares (except for avocados, prunes, raisins, and bananas—none of which come at the top of our list of health-promoters) [15]. He also called for a national campaign (which has failed to materialize). He estimated the potential saving of some $150 billion a year if the U.S. diet could be improved simply to conform to present official recommendations. The National Cancer Institute has started a project in conjunction with the agricultural industry with the aim of increasing people's fruit and vegetable intake.

A paper from South Africa entitled "Public Nutrition: Who Is Listening, Responding, and Acting?" makes even more depressing reading [226]. In it, Walker makes the following points based on a survey of the current literature:

- The majority (90 percent) of the public in the United States (and many other industrialized countries) do not follow official nutritional guidelines.
- There has been no reduction in the intake of calories, total fat, and saturated fat.
- The consumption of fruits and vegetables has not increased.
- Poor people cannot afford enough fruits and vegetables.
- In many developing countries previously healthy diets are being replaced by the injurious Western diet.

- Of seven surveys in various countries of the health effects of the diet during the World War II period, all showed a reduction in obesity, diabetes, coronary heart disease, and dental caries. During that period there was a profound decrease in the consumption of animal fats, sugar, and meat. Consumption of many basic vegetables rose. Obesity and diabetes are currently on the increase among many segments of the world's population.
- Current television advertising by the food industry concentrates on trying to sell food high in fat, sugar, and salt and is particularly aimed at children.

Parents need to redouble their efforts to get their children to eat fruits and vegetables, which, for some unfathomable reason, many of them seem to detest. Furthermore, how many doctors, who treat common chronic diseases, actively encourage their patients and their patients' families to follow this sound nutritional advice? A study in Massachusetts of primary care physicians found that the amount of dietary advice they give is actually declining [227]. The researchers said that this was probably owing to the lack of valid and consistent data presented to support many official dietary recommendations. Other studies of primary care physicians in Australia and Holland have shown that their difficulties arose for several reasons. These physicians felt uncertain as to whether they were entitled to interfere in their patient's "lifestyles" unless asked to do so. They reflected the emphasis in modern medicine on cure rather than prevention and confessed to a lack of knowledge of the field, lack of confidence in their expertise on nutritional issues, and lack of time to discuss these matters during a busy clinical practice.

One potent reason why the public does not take the advice to eat more fruits and vegetables seriously, I suggest, is the failure of the authorities (such as the Food and Nutrition Board of the National Academy of Sciences) to raise the RDAs of antioxidant vitamins and related substances. Advice to the public must be backed up by statements from the National Research Council, the National Academy of Sciences, the FDA, the Surgeon General, and other similar bodies explaining that people should take more antioxidant vitamins than just the minimum amount needed to prevent vitamin deficiency diseases in view of the now overwhelming evidence that antioxidant vitamins, and antioxidants in general, do much more than that. Health officials must also then urge people to obtain their vitamins and antioxidants, as far as is possible, by improving their diets. One group offering health-related pronouncements, the Alliance for Aging Research, has in fact urged that the RDAs be raised immediately.

Conceding that many people will still not follow good advice about diet, we need to advise the next best thing, which is to take, under medical supervision, a balanced supplement regimen containing as wide a variety of synergistic antioxidants as can be obtained. Medical supervision is advisable for these reasons: A person may have a condition such as iron overload, or be a kidney stone former, without knowing it. Antioxidants could interfere with the way the body uses reactive oxygen species in fighting acute bacterial infections. Also, some anticancer drugs depend on reactive oxygen species for their effectiveness. Thus, during the course of an acute bacterial infection or during cancer chemotherapy, antioxidants should be taken only upon medical advice. Finally, a person is well-advised to take a supplement program designed for his or her own particular needs; and only a doctor can determine what these particular needs are. The Alliance for Aging Research recommendations for daily intake levels are vitamin C, 250–1,000 mg; vitamin E, 100–400 mg; and beta-carotene, 10–30 mg. They also recommend that a physician always be consulted to supervise this program.

It has been claimed that benefits from adding trace minerals to the regimen are likely to be minimal as these will simply not be absorbed, except in cases where the patient's diet is so poor as to be short of essential minerals. However, as we have seen, calcium and magnesium form very insoluble salts with undesirable fatty acids in the intestines and thereby help prevent their absorption, thus removing one of the risk factors for colorectal cancer. People at risk for osteoporosis have a special need for an adequate intake of calcium.

The failure of the general population to follow good nutritional advice is paralleled by the inability or unwillingness of the bulk of the population to take enough exercise, as many surveys show, and by the fact that many people still smoke tobacco, the single biggest problem in preventative medicine.

Most participants in this debate assess the need to give or not to give antioxidant supplements from the viewpoint of just one disease, whether heart disease, cancer, cataract, diabetes, and so on. However, raising antioxidant intake may protect against a wide range of diseases, which increases the benefit side of the risk-benefit ratio.

The oxidative damage that eventually leads to chronic diseases has a cumulative effect over many years. It would be inefficient to try to correct this by means of antioxidant intervention at a late stage in the process. The antioxidants should be given before the damage is done, as we saw clearly in the case of type I diabetes and Parkinson's disease. The trouble is that it is difficult to predict who is going to get a particular disease. In certain diseases, such as atherosclerosis and type I diabetes, there are high risk factors we can measure. But what about those diseases, such as Parkinson's disease, no such measures yet exist? If everyone over the age of forty were to change to a Mediterranean-style diet, or partake of an adequate antioxidant program as well as the best diet they could manage, it seems very likely that the incidence of many chronic diseases would fall. Of course, this cannot be absolutely guaranteed, but medicine would be in a sorry state if only mathematical degrees of certainty were to be accepted. William Raspberry, in a column in the *Washington Post*, has predicted that, in view of the current serious financial crisis faced by Medicare, it will soon be fiscally essential for people to adopt a better diet and healthier lifestyle.

Different levels of antioxidant supplementation may be needed for different purposes. For example, Pryor showed that levels of vitamin E well below the RDA level are enough to protect against myopathy (muscle damage) [167]. Prevention of red blood cell hemolysis (rupture) requires the RDA level. In these two cases giving more than the RDA conveys no further benefit. But to improve the function of the immune system, much higher levels than the RDA are required for maximum benefit. As Pryor concludes, "we have passed a watershed with regard to our attitude toward the use of micronutrients and the antioxidant vitamins."

Last, and most important, there are considerable differences among people as to their antioxidant status and requirements. Therefore, suggesting one level of supplementation for everybody is an inefficient way to proceed. Remember the advice given by Tengerdy that, for vitamin E, the optimum dose depends on many factors and must be determined for each individual [213]. This applies to other antioxidants as well. I agree with Snodderly's suggestion that one rational thing to do is to obtain a comprehensive blood-level survey of the most important antioxidants [204]. A good method of doing this is called a Pantox profile, which includes a lipid and iron-balance profile. This test determines which antioxidants a person is taking enough of to meet his or her needs, and of which more or less is needed.

It seems likely that within a few years a physician will no sooner neglect to carry out a plasma antioxidant profile in his patients as fail to do a blood glucose or blood pressure measurement. I suggest that all patients with any of the diseases I have listed, in which oxidative stress plays an important role, ask that such a profile be performed as an essential part of the investigative workup of their illness. Apparently healthy people, too, need a regular monitoring of their antioxidant profile as a measure that may contribute to preventing the development of such diseases later in life. Physicians today expend an enormous amount of time and energy in helping patients fight off attacks by deadly bacteria and viruses. It is now time to focus more attention on helping people fight off the attacks on their bodies by reactive oxygen species, which have the potential to be equally deadly.

The following is a list of fruits and vegetables and other foods that are good sources of antioxidants. A plus sign denotes an especially good source.

CAROTENES

BETA-CAROTENE	apricots, beet greens, cantaloupe, carrots, chicory, collard greens, fennel, kale, mustard greens, parsley, peaches, pumpkin, red pepper, romaine lettuce, spinach, sweet potatoes, Swiss chard, watercress, winter squash
LUTEIN AND ZEAXANTHIN	kale (+), broccoli, spinach, winter squash, Brussels sprouts, celery, dill, leaf lettuce, leeks, mustard greens, parsley, peas, scallions, summer squash
LYCOPENE	tomatoes (+), apricots, guava, pink grapefruit, mango, oranges, peaches, papaya, watermelon
VITAMIN C	broccoli, Brussels sprouts, cauliflower, citrus fruits, green pepper, kiwi, kohlrabi, papaya, peaches, red cabbage, red pepper, strawberries, potatoes, alfalfa, lettuce
VITAMIN E	wheat germ, seeds, nuts, margarine, dark green leafy vegetables, avocado, peanuts, sweet potatoes
LIPOIC ACID	dark green leafy vegetables, especially spinach and broccoli

FLAVONOIDS	apples, citrus fruits, flaxseed, licorice, lentils, onions, peanuts, rice, soy beans, many herbs, blueberries, cranberries (+), black and red currants, loganberries (but low in strawberries and raspberries); chamomile tea, olive oil, red wine, tea
POLYPHENOLS	chocolate, coffee, grapes, nuts, oranges, strawberries, tea, turmeric, white wine
OTHER ANTIOXIDANTS	chives, garlic, shallots, Brussels sprouts, rosemary, yeast

Suggested daily amounts for effective supplements of antioxidants and antioxidant helpers:

CAROTENES

BETA-CAROTENE	10 mg
LUTEINE	10 mg
LYCOPENE	10 mg
ZEAXANTHIN	10 mg

VITAMIN B_3 (NICOTINAMIDE)	100 mg
VITAMIN C	500 mg
VITAMIN E (ALPHA-TOCOPHEROL)	400–800 mg
GAMMA-TOCOPHEROL	100 mg
VITAMIN Q_{10}	300 mg

GLUTAMINE (GLUTATHIONE PRECURSOR)	500 mg
MIXED ISOPRENOIDS	20 mg
SELENIUM	50 µg
ZINC	25 mg

To the above, it would be prudent to add vitamin B_6 (200 mg), B_{12} (1 mg), and folic acid (2 mg), as modern diets are often low in these particular vitamins.

However, it might be wiser to have an antioxidant profile measured by doctor and, following his or her advice, to adjust antioxidant intake in accordance with the profile.

ALS	amytrophic lateral sclerosis
CARET	Beta-Carotene and Retinol Efficiency Trial
CAT	catalase, an antioxidant enzyme
CHAOS	Cambridge Heart Antioxidant Study
DNA	deoxyribonucleic acid, the carrier of genetic information
G	gram
GSHpx	glutathione peroxidase, an antioxidant enzyme
HDL	high-density lipoprotein
L	liter
LDL	low-density lipoprotein
L-DOPA	levo-dihydroxyphenylalanine
MDA	malonyl dialdehyde, a marker for fat oxidation
mg	milligram
NAC	n-acetyl cysteine, an antioxidant
NF-κB	a transcription factor, protein that switches certain genes on and off
NHANES-I	First National Health and Nutrition Examination Survey
PG H	prostaglandin H
PH	Physician's Health
RDA	recommended daily allowance
RNA	ribonucleic acid
SOD	superoxide dismutase, an antioxidant enzyme
TGFα	a measure of inflammation

abbreviations

1. Abrams B, Duncan D, and Hertz-Picciotto I. (1991) A prospective study of dietary intake and acquired immune deficiency syndrome in HIV-seropositive homosexual men. *Journal of Acquired Immune Deficiency Syndrome*, 6, 949–958.

2. Aillet F, Gougerot-Pocidalo M-A, Virelizier J-L et al.(1994) Appraisal of potential therapeutic index of antioxidants on the basis of their in vitro effects on HIV replication in monocytes and interleukin 2-included lymphocyte proliferation. *AIDS Research and Human Retroviruses*, 10, 405–411.

3. Albanes D, Heinonen OP, Huttunen JK et al.(1995) Effects of alpha-tocopherol and beta-carotene supplements on cancer incidence in the Alpha-Tocopherol and Beta-Carotene Cancer Prevention Study. *American Journal of Clinical Nutrition*, 62 (suppl.), 1427S–1430S.

4. Ambrosone CB, Graham S, Marshall JR, et al.(1994) Dietary antioxidants and breast cancer risk: effect modification by family history. *Advances in Experimental Biology and Medicine*, 366, 439–440.

5. The Alpha-Tocopherol, Beta-Carotene Cancer Prevention Study Group. (1994) The effect of vitamin E and beta-carotene on the incidence of lung cancer and other cancers in male smokers. *New England Journal of Medicine*, 330, 1029–1035.

6. Armstrong AM, Chestnutt JE, Gormley MJ et al. (1996) The effect of dietary treatment on lipid peroxidation and antioxidant status in newly diagnosed noninsulin dependent diabetes. *Free Radical Biology and Medicine*, 21, 719–726.

7. Azhar S, Cao L., and Reaven E. (1995) Alteration of the adrenal antioxidant defense system during aging in rats. *Journal of Clinical Investigation*, 96, 1414–1424.

8. Bao-Khanh Q, Do BS, Garewal HS, Clemens NC Jr. et al. (1996) Exhaled ethane and antioxidant vitamin supplements in active smokers. *Chest*, 110, 159–164.

9. Barber DA and Harris SR. (1994) Oxygen free radicals and antioxidants: a review. *American Pharmacist*, NS34, 26–35.

references

10. Barton JC and Bertoli LF. (1996) Hemachromatosis: the genetic disorder of the 21st century. *Nature (Medicine)*, 2, 394–395.

11. Bendich A. (1996) Antioxidant vitamins and human immune responses. *Vitamins and Hormones*, 52, 35–62.

12. Bendich A and Langseth L. (1995) The health effects of vitamin C supplementation: a review. *Journal of the American College of Nutrition*, 14, 124–136.

13. Bernard GR, Wheeler AP, Arons MM et al. (1997) A trial of antioxidants N-acetylcysteine and procysteine in ARDS. *Chest*, 112, 164–172.

14. Bettger WJ. (1993) Zinc and selenium, site-specific versus general antioxidation. *Canadian Journal of Physiology and Pharmacology*, 71, 721–724.

15. Blackburn GL. (1992) Improving the American diet. *American Journal of Public Health*, 82, 465–466.

16. Block G. (1995) Are clinical trials really the answer? *American Journal of Clinical Nutrition*, 62 (suppl.), 1517S–1520S.

17. Block G, Patterson B, and Subar A. (1992) Fruit, vegetables, and cancer prevention: a review of the epidemiological evidence. *Nutrition and Cancer*, 18, 1–29.

18. Blot WJ, Li J-Y, Taylor PR et al. (1993) Nutrition intervention trials in Linxian, China: supplementation with specific vitamin/mineral combinations, cancer incidence, and disease-specific mortality in the general population. *Journal of the National Cancer Institute*, 85, 1483–1492.

19. Blumberg J. (1994) Are antioxidants at an awkward age? *Journal of the American College of Nutrition*, 13, 218–219.

20. Böhm F, Edge R, Land EJ et al. (1997) Carotenoids enhance vitamin E antioxidant efficiency. *Journal of the American Chemical Society*, 119, 621–627.

21. Bonithon-Kopp C, Coudray C, Berr C et al. (1997) Combined effects of lipid peroxidation and antioxidant status on carotid atherosclerosis in a population aged 59–71 years: the EVA Study. *American Journal of Clinical Nutrition*, 65, 121–127.

22. Brown KM, Morrice PC, Arthur JR et al. (1996) Effects of vitamin E supplementation on erythrocyte antioxidant defence mechanisms in smoking and non-smoking men. *Clinical Science*, 91, 107–111.

23. Browne SE, Bowling AC, MacGarvey W et al. (1997) Oxidative damage and metabolic dysfunction in Huntington's disease: selective vulnerability of the basal ganglia. *Annals of Neurology*, 41, 646–653.

24. Bukin YV and Draudin-Krylenko VA. (1997) The role of β-carotene and vitamin E in the treatment of gastric premalignant lesions: biochemical and clinical aspects. *Abstracts. VIth World Congress of Nutrition*, 29.

25. Bussey HJR, De Cosse JJ, Deschner EE et al. (1982) A randomized trial of ascorbic acid in polyposis coli. *Cancer*, 50, 1434–1439.

26. Byers T and Bowman B. (1993) Vitamin E supplements and coronary heart disease. *Nutrition Reviews*, 51, 333–336.

27. Cals MJ, Succari M, Meneguzzer E et al. (1997) Markers of oxidative stress in fit, health-conscious elderly people living in the Paris area. *Nutrition*, 13, 319–326.

28. Carotenoid Research Interactive Group. (1996) Beta-carotene and the carotenoids: beyond the intervention trials. *Nutrition Reviews*, 54(6), 185–187.

29. Ceballos-Picot I, Witko-Sarsat V, Merad-Boudia M et al. (1996) Glutathione antioxidant system as a marker of oxidative stress in chronic renal failure. *Free Radical Biology and Medicine*, 21, 845–853.

30. Chalmers AH, Cowley DM, and Brown JM. (1986) A possible etiological role for ascorbate in calculi formation. *Clinical Chemistry*, 32, 333–336.

31. Chamiec T, Herbaczynska-Cedro K, and Ceremuzynski L. (1996) Effects of antioxidant vitamins C and E on signal-averaged electrocardiogram in acute myocardial infarction. *American Journal of Cardiology*, 77, 237–241.

32. Chandrasekar B, Colston JT, and Freeman GL. (1997) Induction of proinflammatory cytokine and antioxidant enzyme gene expression following brief myocardial ischemia. *Clinical and Experimental Immunology*, 108, 346–351.

33. Chase HP, Butler-Simon N, Garg S et al. (1990) A trial of nicotinamide in newly diagnosed patients with type 1 (insulin-dependent) diabetes mellitus. *Diabetologia*, 33, 444–446.

34. Christen WG Jr. (1994) Antioxidants and eye disease. *American Journal of Medicine*, 97 (Suppl. 3A), 14S–17S.

35. Comstock GW, Helzlsouer KJ, and Bush TL. (1991) Prediagnostic serum levels of carotenoids and vitamin E as related to subsequent cancer in Washington County, Maryland. *American Journal of Clinical Nutrition*, 53 (suppl.), 260S–264S.

36. Cook NC and Samman S. (1996) Flavonoids—chemistry, metabolism, cardioprotective effects, and dietary sources. *Nutritional Biochemistry*, 7, 66–76.

37. Daily JW and Zemel MB. (1995) Antioxidants: research vs rhetoric. *American Journal of Clinical Nutrition*, 61, 866–868.

38. Darr D, Combs S, and Pinnell, S. (1993) Ascorbic acid and collagen synthesis: rethinking a role for lipid peroxidation. *Archives of Biochemistry and Biophysics*, 307, 331–335.

39. Das D, Bandyopadhyay D, Bhattacherjee M et al. (1997) Hydroxyl radical is the major causative factor in stress-induced gastric ulceration. *Free Radical Biology and Medicine*, 23, 8–18.

40. Delafuente JC, Prendergast JM, and Modigh A (1986) Immunological modulation by vitamin C in the elderly. *International Journal of Immunopharmacology*, 8, 205–221.

41. de Rijk MC, Breteler MBB, den Breeijen JH et al. (1997) Dietary antioxidants and Parkinsons disease: the Rotterdam study. *Archives of Neurology*, 54, 762–765.

42. de Rijke YB, Demacker PNM, Assen NA et al. (1995) Red wine consumption does not affect oxidizability of low-density lipoproteins in volunteers. *American Journal of Clinical Nutrition*, 63, 329–334.

43. Diplock AT. (1994) Antioxidants and disease prevention. *Molecular Aspects of Medicine*, 15, 293–376.

44. Diplock AT. (1995) Safety of antioxidant vitamins and β-carotene. *American Journal of Clinical Nutrition*, 62 (suppl.), 1510S–1516S.

45. Draper HH and Bettger WJ. (1994) Role of nutrients in the cause and prevention of oxygen radical pathology. *Advances in Experimental Medicine and Biology*, 366, 269–289.

46. Dröge W, Gross A, Hack V et al. (1997) Role of cysteine and glutathione in HIV infection and cancer cachexia: therapeutic intervention with N-acetylcysteine. *Advances in Pharmacology*, 38, 581–600.

47. Dröge W, Holm E. (1997) Role of cysteine and glutatione in HIV infection and other diseases associated with muscle wasting and immunological dysfunction. *FASEB Journal*, 11, 1077–1089.

48. Drukarch B, Langeveld CH, and Stoof JC. (1997) Glutathione homeostasis is linked to the vesicular storage of dopamine in rat PC12 phechromocytoma cells. *Experimental Neurology*, 145 (suppl.), S39.

49. Edwards CQ and Kushner JP. (1993) Screening for hemochromatosis. *New England Journal of Medicine*, 328, 1616–1620.

50. Efendy JL, Simmons DL, and Campbell GR et al. (1997) The effect of the aged garlic extract, "Kyolic," on the development of experimental atherosclerosis. *Atherosclerosis*, 132, 37–42.

51. Enstrom JE, Kanim LE, and Klein MA. (1992) Vitamin C intake amd mortality amongst a sample of the United States population. *Epidemiology*, 3, 194–202.

52. Enstrom JE. (1993) Counterviewpoint: vitamin C and mortality. *Nutrition Today*, 28 (3), 39–42.

53. Evans PH. (1993) Free radicals in brain metabolism and pathology. *British Medical Bulletin*, 49, 577–587.

54. Everall IP, Hudson L, and Kerwin RW. (1997) Decreased absolute levels of ascorbic acid and unaltered vasoactive intestinal polypeptide receptor binding in the frontal cortex in acquired immunodeficiency syndrome. *Neuroscience Letters*, 224, 119–122.

55. Flagg EW, Coates RJ, and Greenberg RS. (1995) Epidemiologic studies of antioxidants and cancer in humans. *Journal of the American College of Nutrition*, 14, 419–427.

56. Floren LC, Zangwill AC, and Schroeder DJ. (1994) Antioxidants may retard cataract formation. *Annals of Pharmacotherapy*, 28, 1040–1042.

57. Franke AA, Harwood PJ, Shimamoto T et al. (1994) Effects of micronutrients and antioxidants on lipid peroxidation in human plasma and in cell culture. *Cancer Letters*, 79, 17–26.

58. Freudenheim JL, Marshall JR, Vena JE et al. (1996) Premenopausal breast cancer risk and intake of vegetables, fruits and related nutrients. *Journal of the National Cancer Institute*, 88, 340–348.

59. Fuhrman B, Lavy A, and Aviram M. (1995) Consumption of red wine with meals reduces the susceptibility of human plasma and low-density lipoprotein to lipid peroxidation. *American Journal of Clinical Nutrition*, 61, 549–554.

60. Fujiki H, Suganuma M, Okabe S et al. (1996) Japanese green tea as a cancer preventive in man. *Nutrition Reviews*, 54 (suppl.), S67–S70.

61. Galley HF, Thornton J, Howdle PD et al. (1997) Combination oral antioxidant supplementation reduces blood pressure. *Clinical Science*, 92, 361–365.

62. Garewal HS and Diplock AT. (1995) How "safe" are antioxidant vitamins? *Drug Safety*, 13, 8–14.

63. Gershoff SN. (1993) Vitamin C (ascorbic acid): new roles, new requirements? *Nutrition Reviews*, 51, 313–326.

64. Gey KF. (1993) Prospects for the prevention of free radical disease, regarding cancer and cardiovascular disease. *British Medical Bulletin*, 49, 679–699.

65. Gey KF. (1994) Optimum plasma levels of antioxidant micronutrients. Ten years of antioxidant hypothesis on arteriosclerosis. *Biblioteca Nutritio et Dieta*, 51, 84–99.

66. Gey KF, Stähelin HB, and Eichholzer M. (1993) Poor plasma status of carotene and vitamin C is associated with higher mortality from ischaemic heart disease and stroke: Basel Prospective Study. *Clinical Investigator*, 71, 3–6.

67. Giavannuchi E, Ascherio A, Rimm EB et al. (1995) Intake of carotenoids and retinol in relation to risk of prostate cancer. *Journal of the National Cancer Institute*, 87, 1767–1776.

68. Gillissen A, Jaworska M, Schärling B et al. (1997) Beta-2-agonists have antioxidant functions in vitro. *Respiration*, 64, 16–22.

69. Goldberg DM. (1995) Editorial: does wine work? *Clinical Chemistry*, 41, 14–16.

70. Goldberg DM. (1996) More on antioxidant activity of reversatol in red wine. *Clinical Chemistry*, 42, 113–114.

71. Goldbohm RA, Verhoeven DTH, Voorrips LE et al. (1997) Consumption of fruit and vegetables and risk of lung cancer in a prospective study. *American Journal of Epidemiology*, 145, S71.

72. Goldfarb S. (1994) Diet and nephrolithiasis. *Annual Review of Medicine*, 45, 235–243.
73. Good PF, Werner P, Hsu A et al. (1996) Evidence for neuronal oxidative damage in Alzheimer's disease. *American Journal of Pathology*, 149, 21–28.
74. Greenberg ER and Sporn MB. (1996) Editorial: antioxidant vitamins, cancer and cardiovascular disease. *New England Journal of Medicine*, 334, 1189–1190.
75. Gridley G, McLaughlin JK, Block G et al. (1992) Vitamin supplement use and reduced risk of oral and pharyngeal cancer. *American Journal of Epidemiology*, 135, 1083–1092.
76. Haber B. (1997) The Mediterranean diet: a view from history. *American Journal of Clinical Nutrition*, 665 (suppl.), 1053S–1057S.
77. Hankinson SE, Stampfer MJ, Seddon JM et al. (1992) Nutrient intake and cataract extraction in women: a prospective study. *British Medical Journal*, 305, 335–339.
78. Hannon MP, Hughes C, O'Kane MJ et al. (1997) Oxidative stress and DNA damage in platelets in patients with insulin-dependent diabetes mellitus. *Abstracts. VIth World Congress of Clinical Nutrition*, 35.
79. Hatch GE. (1995) Asthma, inhaled oxidants, and dietary antioxidants. *American Journal of Clinical Nutrition*, 61 (suppl.), 625S–630S.
80. Hathcock JN. (1997) Vitamins and minerals: efficacy and safety. *American Journal of Clinical Nutrition*, 66, 427–437.
81. Hemilä H and Herman ZS. (1996) Vitamin C and the common cold: a retrospective study of Chalmers' review. *American Journal of Clinical Nutrition*, 14, 116–123.
82. Hennekens CH, Buring JE, Manson JE et al. (1996) Lack of effect of long-term supplementation with beta carotene on the incidence of malignant neoplasms and cardiovascular disease. *New England Journal of Medicine*, 334, 1145–1149.
83. Herbert V. (1993) Viewpoint. Does mega-C do more good than harm, or more harm than good? *Nutrition Today*, 28, 28–32.
84. Herbert V. (1994) The antioxidant supplement myth. *American Journal of Clinical Nutrition*, 60, 157–158.
85. Herbert V. (1995) Editorial. Vitamin C supplements and disease—counterpoint. *Journal of the American College of Nutrition*, 14, 112–113.
86. Hertog MGL, Feskens EJM, Holman PCH et al. (1993) Dietary antioxidant flavonoids and risk of coronary heart disease: the Zutphen Elderly Study. *Lancet*, 342, 1007–1011.
87. Hertog MGL, Sweetman PM, Fehily AM et al. (1997) Antioxidant flavonols and ischemic heart disease in a Welsh population of men: the Caerphilly study. *American Journal of Clinical Nutrition*, 65, 1489–1494.
88. Hodis HN, Mack WJ and LaBree L. (1995) Serial coronary angiographic evidence that antioxidant vitamin intake reduces progression of coronary artery atherosclerosis. *Journal of the American Medical Association*, 273, 1845–1854.
89. Hoffer A, Osmond H, and Smythies J. (1954) Schizophrenia. A new approach. Part II. *Journal of Mental Science*, 100, 29–37.
90. Hoffman RM and Garewal HS. (1995) Antioxidants and the prevention of coronary heart disease. *Archives of Internal Medicine*, 155, 241–246.
91. Hollman PCH, Hertog MGL, and Katan MB. (1996) Role of dietary flavonoids in protection against cancer and coronary heart disease. *Biochemical Society Transactions*, 24, 785–789.
92. Hu G, Cassano P, and Chen J. (1997) Dietary vitamin C intake and lung function in rural China. *American Journal of Epidemiology*, 145 (suppl.), S80.
93. Hubel CA, Kagan VE, Kisin ER et al. (1997) Increased ascorbate radical formation and ascorbate depletion in plasma from women with preeclampsia: implications for oxidative stress. *Free Radical Biology and Medicine*, 23, 597–609.

94. Illingworth DR. (1993) The potential role of antioxidants in the prevention of athero-sclerosis. *Journal of Nutritional Science and Vitaminology*, 39 (suppl.), S43–S47.

95. Ip C, Lisk DJ, and Scimeca JA. (1994) Potential of food modification in cancer pre-vention. *Cancer Research*, 54 (suppl.), S1957–S1959.

96. Jacques PF, Halpner AD, and Blumberg JB. (1995) Influence of combined antioxidant nutrient intakes on their plasma concentrations in an elderly population. *American Journal of Clinical Nutrition*, 62, 1228–1233.

97. Jacques PF, Taylor A, Hankinson SE et al. (1997) Long-term vitamin C supplement use and prevalence of early age-related lens opacities. *American Journal of Clinical Nutrition*, 66, 911–916.

98. Johnson LE. (1994) The emerging role of vitamins as antioxidants. *Archives of Family Medicine*, 3, 809–820.

99. Kampman E, Verhoeven D, Sloots L et al. (1995) Vegetable and animal products as determinants of colon cancer risk in Dutch men and women. *Cancer Causes and Control*, 6, 225–234.

100. Kanofsky JD and Sandyk R. (1992) Antioxidants in the treatment of schizophrenia. *International Journal of Neuroscience*, 62, 97–100.

101. Kaugars GE, Silverman S Jr., Lovas JGL et al. (1993) A review of the use of anti-oxidant supplements in the treatment of human oral leukoplakia. *Journal of Cellular Biochemistry*, 17F (suppl.), 292–298.

102. Keimowitz RM. (1997) Dementia improvement with cytotoxic chemotherapy: a case of Alzheimer's disease and multiple myeloma. *Archives of Neurology*, 54, 485–488.

103. Kennes B, Dumont I, Brohee D et al. (1983) Effect of vitamin C supplements on cell-mediated immunity in old people. *Gerontology*, 29, 305–310.

104. Kimmick GG, Bell RA, and Bostick RM. (1997) Vitamin E and breast cancer: a review. *Nutrition and Cancer*, 27, 109–117.

105. Knauf VC and Facciotti D. (1995) Genetic engineering of foods to reduce the risk of heart disease and cancer. *Advances in Experimental Biology and Medicine*, 369, 221–228.

106. Knekt P, Heliövaara M, Rissanen A et al. (1992) Serum antioxidant vitamins and risk of cataract. *British Medical Journal*, 305, 1392–1394.

107. Knekt P, Järvinen R, Reunanen A et al. (1996) Flavonoid intake and coronary mor-tality in Finland: a cohort study. *British Medical Journal*, 312, 478–481.

108. Knekt P, Järvinen R, Seppänen R et al. (1997) Dietary flavonoids and the risk of lung cancer and other malignant neoplasms. *American Journal of Epidemiology*, 146, 223–230.

109. Koedel W and Pfister H-W. (1997) Protective effect of the antioxidant N-acetyl-L-cysteine in pneumococcal meningitis in the rat. *Neuroscience Letters*, 225, 33–36.

110. Kohlmeier L, Kark JD, Gomez-Garcia E et al. (1997) Lycopene and myocardial infarc-tion risk in the EURAMIC Study. *American Journal of Epidemiology*, 146, 618–626.

111. Kristenson M, Ziedén B, Kucinskienë Z et al. (1997) Antioxidant state and mortality from coronary heart disease in Lithuanian and Swedish men: concomitant cross-sectional study of men aged 50. *British Medical Journal*, 314, 629–633.

112. Kritchevsky SB, Schwartz GG, and Morris DL. (1995) Beta-carotene supplementation, vitamin D, and cancer risk: a hypothesis. *Epidemiology*, 6, 89.

113. Kuklinski B, Weissenbacher E, and Fähnrich A. (1994) Coenzyme Q_{10} and anti-oxidants in acute myocardial infarction. *Molecular Aspects of Medicine*, 15 (suppl.), S143–S147.

114. Kushi LH, Folsom AR, Prineas RJ et al. (1996) Dietary antioxidant vitamins and death from coronary heart disease in postmenopausal women. *New England Journal of Medicine*, 334, 1156–1162.

115. Landrum JT, Bone RA, and Kilburn MD. (1997) The macular pigment: possible role in protection from age-related macular degeneration. *Advances in Pharmacology*, 38, 537–556.

116. Langsjoen H, Langsjoen P, Langsjoen P et al. (1994a) Usefulness of coenzyme Q_{10} in clinical cardiology: a long term study. *Molecular Aspects of Medicine*, 15 (suppl.), S165–S175.

117. Langsjoen P, Langsjoen P, Willis R et al. (1994b) Treatment of essential hypertension with coenzyme Q_{10}. *Molecular Aspects of Medicine*, 15 (suppl.), S265–S272.

118. Leeuwenburgh C, Fiebig R, Chandwaney R et al. (1995) Aging and exercise training in skeletal muscle: responses of glutathione and antioxidant enzyme systems. *American Journal of Physiology*, 267, R439–R445.

119. Lehr H-A and Messmer K. (1996) Rationale for the use of antioxidant vitamins in clinical organ transplantation. *Transplantation*, 62, 1197–1199.

120. Lehr, H-A, Frei B, and Arfors K-E. (1994) Vitamin C prevents cigarette smoke-induced leukocyte aggregation and adhesion to endothelium in vivo. *Proceedings of the National Academy of Sciences. USA*, 91, 7688–7692.

121. Leske MC, Chylack LT Jr, and Wu SY. (1991) The Lens Opacities Case-Control Study. Risk factors for cataract. *Archives of Ophthalmology*, 109, 244–251.

122. Levine M, Dhariwal KR, Welch RW et al. (1995) Determination of optimal vitamin C requirements in humans. *American Journal of Clinical Nutrition*, 62 (suppl.), S1347–S1356.

123. Levine M, Conry-Cantilena C, Wang Y et al. (1996) Vitamin C pharmacokinetics in healthy volunteers: evidence for a recommended daily allowance. *Proceedings of the National Academy of Sciences. USA*, 93, 3704–3709.

124. Lih-Brody L, Powell SR, Collier KP et al. (1996) Increased oxidative stress and decreased antioxidant defenses in mucosa of inflammatory bowel disease. *Digestive Diseases and Sciences*, 41, 2078–2086.

125. Liu J and Mori A. (1993) Monoamine metabolism provides an antioxidant defense in the brain against oxidant- and free radical-induced damage. *Archives of Biochemistry and Biophysics*, 302, 118–127.

126. Livrea MA, Tesoriere L, Pintaudi AM et al. (1996) Oxidative stress and antioxidant status in β-thalassemia major: iron overload and depletion of lipid soluble antioxidants. *Blood*, 88, 3608–3614.

127. Lockwood K, Moesgaard S, Hanioka T et al. (1994) Apparent partial remission of breast cancer in "high risk" patients supplemented with nutritional antioxidants, essential fatty acids and coenzyme Q_{10}. *Molecular Aspects of Medicine*, 15, S231–S240.

128. Losonczy KG, Harris TB, and Havlik RJ. (1996) Vitamin E and vitamin C supplement use and risk of all-cause and coronary heart disease mortality in older persons: the Established Populations for Epidemiological Studies of the Elderly. *American Journal of Clinical Nutrition*, 64, 190–196.

129. Luoma PV, Näyhä S, Sikkila K et al. (1995) High serum alpha-tocopherol, albumin, selenium and cholesterol, and low mortality from coronary heart disease in northern Finland. *Journal of Internal Medicine*, 237, 49–54.

130. Lykkesfeldt J, Loft S, Nielsen JB et al. (1997) Ascorbic acid and dehydroascorbic acid as biomarkers of oxidative stress caused by smoking. *American Journal of Clinical Nutrition*, 65, 959–963.

131. Mares-Perlman JA, Brady WE, Klein R et al. (1995) Serum antioxidants and age-related macular degeneration in a population-based case-control study. *Archives of Ophthalmology*, 113, 1518–1523.

132. Maxwell SRJ. (1995) Prospects for the use of antioxidant therapies. *Drugs*, 49, 345–361.

133. Maxwell SRJ, Thomason H, Sandler D et al. (1997) Antioxidant status in patients with uncomplicated insulin-dependent and non-insulin-dependent diabetes mellitus. *European Journal of Clinical Investigation*, 27, 484–490.

134. McGeer PL and McGeer EG. (1995) The inflammatory response system of brain: implications for therapy of Alzheimer's and other neurodegenerative diseases. *Brain Research Reviews*, 21, 195–218.

135. McGeer PL, Schulzer M, and McGeer EG. (1996). Arthritis and anti-inflammatory agents as possible protective factors for Alzheimer's disease: a review of 17 epidemiological studies. *Neurology*, 47, 425–432.

136. McKeown-Eyssen G, Holloway C, Jazmaji V et al. (1988) A randomized trial of vitamins C and E in the prevention of recurrence of colorectal polyps. *Cancer Research*, 48, 4701–4705.

137. Medina JH, Viola H, Wolfman C et al. (1997) Overview—flavonoids: a new family of benzodizepine receptor ligands. *Neurochemistry Research*, 22, 419–425.

138. Mehra MR, Lavie CJ, Ventura HO et al. (1995) Prevention of atherosclerosis. The potential role of antioxidants. *Postgraduate Medicine*, 98, 175–184.

139. Mehta J. (1997) Intake of antioxidants among American cardiologists. *American Journal of Cardiology*, 79, 1558–1560.

140. Mena MA, Pardo B, Paino CL et al. (1993) Levodopa toxicity in foetal rat midbrain neurones in culture: modulation by ascorbic acid. *NeuroReport*, 4, 438–440.

141. Meydani SN, Barklund MP, Liu S et al. (1990) Vitamin E supplementation enhances cell-mediated immunity in healthy elderly subjects. *American Journal of Clinical Nutrition*, 52, 557–563.

142. Meyers DG and Maloley PA. (1993) The antioxidant vitamins: impact on atherosclerosis. *Pharmacotherapy*, 13, 574–582.

143. Meyers DG, Maloley PA, and Weeks D. (1996) Safety of antioxidant vitamins. *Archives of Internal Medicine*, 156, 925–935.

144. Miller NJ, Johnston JD, Collis CS et al. (1997) Serum total antioxidant activity after myocardial infarction. *Annals of Clinical Biochemistry*, 34, 85–90.

145. Moertel CG, Fleming TR, Creagan ET et al. (1985) High-dose vitamin C versus placebo in the treatment of patients with advanced cancer who have had no prior chemotherapy. A randomized double-blind comparison. *New England Journal of Medicine*, 312, 137–141.

146. Morris DL, Kritchevsky SB, and Davis CE. (1994) Serum carotenoids and coronary heart disease. *Journal of the American Medical Association*, 272, 1439–1441.

147. Mukhopadhyay CK and Chatterjee B. (1994) NADPH-initiated cytochrome P450-mediated free metal ion-dependent oxidative damage of microsomal proteins: exclusive prevention by ascorbic acid. *Journal of Biological Chemistry*, 269, 13,390–13,397.

148. Nath R, Pandav S, Parashar S et al. (1997) Role of oxygen free radicals in etiopathogenesis of age related macular degeneration (AMD). *Abstracts. VIth World Congress of Clinical Nutrition*, 43.

149. Nyyssönen K, Parviainen M, Salonen R et al. (1997) Vitamin C deficiency and risk of myocardial infarction: prospective population study of men from eastern Finland. *British Medical Journal*, 314, 634–638.

150. Oberley TD and Oberley LW. (1997) Antioxidant enzyme levels in cancer. *Histology and Histopathology*, 12, 525–535.

151. Oliver MF. (1995) Antioxidant nutrients, atherosclerosis, and coronary heart disease. *British Heart Journal*, 73, 299–301.

152. Omaye ST, Burri BJ, Swendseid ME et al. (1996) Blood antioxidants change in young women following β-carotene depletion and repletion. *Journal of the American College of Nutrition*, 15, 469–474.

153. Omenn GS. (1995) Editorial. What accounts for the association of vegetables and fruit with lower incidence of cancers and coronary heart disease? *Annals of Epidemiology*, 5, 333–335.

154. Omenn GS, Goodman GE, Thornquist MD et al. (1996) Effects of a combination of beta-carotene and vitamin A on lung cancer and cardiovascular disease. *New England Journal of Medicine*, 334, 1150–1155.

155. Packer L and Diplock A. *Handbook of Antioxidants*. New York, Dekker, 1996.

156. Packer L, Witt EH, and Tritschler HJ. (1995) Alpha-lipoic acid as a biological antioxidant. *Free Radical Biolology and Medicine*, 19, 227–250.

157. Paller MS and Eaton JW. (1995) Hazards of antioxidant combinations containing superoxide dismutase. *Free Radical Biology and Medicine*, 18, 883–890.

158. Pandey DK, Shekelle R, Selwyn BJ et al. (1996) Dietary vitamin C and beta-carotene and risk of death in middle-aged men: the Western Electric study. *American Journal of Epidemiology*, 142, 1269–1278.

159. Panetta JA, Shadle JK, Phillips ML et al. (1993) 4-thiazolidinones, potent antioxidants as anti-inflammatory agents. *Annals of the New York Academy of Sciences*, 696, 415–416.

160. Paolisso G, Balbi V, Volpe C et al. (1995) Metabolic benefits deriving from chronic vitamin C supplementation in aged non-insulin-dependent diabetics. *Journal of the American College of Nutrition*, 14, 387–392.

161. Parfitt VJ, Rubba P, Bolton C et al. (1994) A comparison of antioxidant status and free radical peroxidation of plasma lipoproteins in healthy young persons from Naples and Bristol. *European Heart Journal*, 15, 871–876.

162. Patterson BH, Block G, Rosenberger WF et al. (1990) Fruits and vegetables in the American diet: data from the NHANES II survey. *American Journal of Public Health*, 80, 1443–1449.

163. Peterson PL. (1995) The treament of mitochondrial myopathies and encephalo-myopathies. *Biochimica et Biophysica Acta*, 1271, 275–280.

164. Poranan AK, Ekblad U, Uotila P et al. (1997) Lipid peroxidation and antioxidants in normal and pre-eclamptic pregnancies. *Placenta*, 17, 401–405.

165. Prasad KN and Kumar R. (1996) Effect of individual and multiple antioxidant vitamins on growth and morphology of human nontumorigenic and tumorigenic parotid acinar cells in culture. *Nutrition and Cancer*, 26, 11–19.

166. Pricmé H, Loft S, Nyyssönen K et al. (1997) No effect of supplementation with vitamin E, ascorbic acid or coenzyme Q_{10} on oxidative NDA damage estimated by 8-oxo-7,8-dihydro-2'-deoxyguanosine excretion in smokers. *American Journal of Clinical Nutrition*, 65, 503–507.

167. Pryor WA. (1991) The antioxidant nutrients and disease prevention—what do we know and what do we need to find out? *American Journal of Clinical Nutrition*, 53 (suppl.), S391–S393.

168. Rachmilewitz EA, Shifter A, and Kahane I. (1979) Vitamin E deficiency in β-thalassemia major: changes in hematological and biochemical parameters after a therapeutic trial with alpha-tocopherol. *American Journal of Clinical Nutrition*, 32, 1850–1858.

169. Ramsey BW, Farrell PM and Pencharz P. (1992) Nutritional assessment and management in cystic fibrosis: a consensus report. *American Journal of Clinical Nutrition*, 55, 108–116.

170. Rapola JM, Virtamo J, Ripatti S et al. (1997) Randomized trial of alpha-tocopherol and beta-carotene supplements on incidence of major coronary events in men with previous myocardial infarction. *Lancet*, 349, 1715–1720.

171. Rautalahti M and Huttunen J. (1994) Antioxidants and carcinogenesis. *Annals of Medicine*, 26, 435–441.

172. Reece EA and Wu Y-K. (1997) Prevention of diabetic embryopathy in offspring of diabetic rats with use of a cocktail of deficient substances and an antioxidant. *American Journal of Obstetrics and Gynecology*, 176, 790–797.

173. Reider CR and Paulson GW. (1997) Lou Gehrig and amyotrophic lateral sclerosis. *Archives of Neurology*, 54, 527–528.

174. Riemersma RA. (1994) Epidemiology and the role of antioxidants in preventing coronary heart disease: a brief overview. *Proceedings of the Nutrition Society*, 53, 59–65.

175. Riemersma RA, Oliver M, Elton RA et al. (1990) Plasma antioxidants and coronary heart disease: vitamins C and E, and selenium. *European Journal of Clinical Nutrition*, 44, 143–150.

176. Rimm EB, Stampfer MJ, Ascherio A et al. (1993) Vitamin E consumption and the risk of coronary disease in men. *New England Journal of Medicine*, 328, 1450–1456.

177. Rivers JM. (1987) Safety of high-level vitamin C ingestion. *Annals of the New York Academy of Sciences*, 498, 445–454.

178. Robertson JM, Donner AP, and Trevithick JR. (1989) Vitamin E intake and risk of cataracts in humans. *Annals of the New York Academy of Sciences*, 570, 372–382.

179. Rodgers AB, Kessler LG, Portnoy LB et al. (1994) "Eat for Health": a supermarket intervention for nutrition and cancer risk reduction. *American Journal of Public Health*, 84, 72–76.

180. Rojas C, Cadenas S, Lopéz-Torres M et al. (1996) Increase in heart glutathione redox ratio and total antioxidant capacity and decrease in lipid peroxidation after vitamin E dietary supplementation in guinea-pigs. *Free Radical Biology and Medicine*, 21, 907–915.

181. Rondanelli M, Melzi d'Eril GV, Anesi A et al. (1997) Altered oxidative stress in healthy old subjects. *Aging. Clinical and Experimental Research*, 9, 221–223.

182. Rucker RB and Stites T. (1994) New perspectives on function of vitamins. *Nutrition*, 10, 507–513.

183. Sano M, Ernesto C, Thomas RG et al. (1997) A controlled study of selegiline, alpha-tocopherol, or both as treatment for Alzheimer's disease. *New England Journal of Medicine*, 336, 1216–1222.

184. Santini SA, Marra G, Giardina B et al. (1997) Defective plasma antioxidant defenses and enhanced susceptibility to lipid peroxidation in uncomplicated IDDM. *Diabetes*, 46, 1853–1858.

185. Sathiyaraj D, Muthu R, and Mohamed AJ. (1997) Antioxidant status, proinflammatory cytokines and acute-phase protein response in smokers after myocardial infarction. *Journal of Clinical and Biochemical Nutrition*, 22, 131–137.

186. Sawyer MAJ, Mike JJ, Chavin K et al. (1989) Antioxidant therapy and survival in ARDS. *Critical Care Medicine*, 17, S153.

187. Schalch W and Weber P. (1994) Vitamins and carotenoids—a promising approach to reducing the risk of coronary heart disease, cancer and eye diseases. *Advances in Experimental Medicine and Biology*, 366, 335–350.

188. Scheider WL, Hershey LA, Vena JE et al. (1997) Dietary antioxidants and other dietary factors in the etiology of Parkinson's disease. *Movement Disorders*, 12, 190–196.

189. Schmidt K-H, Hagmaier V, Hornig DH et al. (1981) Urinary oxalate excretion after large intakes of ascorbic acid in man. *American Journal of Clinical Nutrition*, 34, 305–311.

190. Schünemann HJ, Freudenheim JL, Muti P et al. (1997) Cross-sectional study of vitamin C and a marker for lipid peroxidation. *American Journal of Epidemiology*, 145 (suppl.), S71.

191. Schwartz JC and Shklar G. (1997) Retinoid and carotenoid angiogenesis: a possible explanation for enhanced oral carcinogenesis. *Nutrition and Cancer*, 27, 192–199.

192. Seddon JM, Ajani UA, Sperduto RD et al. (1994) Dietary carotenoids, vitamins A, C, and E, and advanced age-related macular degeneration. *Journal of the American Medical Association*, 272, 1413–1420.

193. Sen CK. (1995) Oxidants and antioxidants in exercise. *Journal of Applied Physiology*, 79, 675–686.

194. Shklar G, Schwartz J, Trickler D et al. (1993) The effectiveness of a mixture of beta-carotene, alpha-tocopherol, glutathione, and ascorbic acid for cancer prevention. *Nutrition and Cancer*, 20, 145–151.

195. Simonian NA and Coyle JT. (1996) Oxidative stress in neurodegenerative diseases. *Annual Review of Pharmacology and Toxicology*, 36, 83–106.

196. Singh RB, Niaz MA, Bishnoi I et al. (1994a) Diet, antioxidant vitamins, oxidative stress and risk of coronary artery disease: the Peerzada Prospective Study. *Acta Cardiologica*, 49, 453–467.

197. Singh RB, Niaz MA, Sharma JP et al. (1994b) Plasma levels of antioxidant vitamins and oxidative stress in patients with acute myocardial infarction. *Acta Cardiologica*, 49, 441–452.

198. Singh RB, Niaz MA, Rastosi SS et al. (1996) Usefulness of antioxidant vitamins in suspected acute myocardial infarction (The Indian Experiment in Infarct Survival-3). *American Journal of Cardiology*, 77, 232–236.

199. Sisto T, Paajanen H, Metsä-Ketelä T et al. (1995) Pretreatment with antioxidants and allopurinol diminishes cardiac onset events in coronary bypass grafting. *Annals of Thoracic Surgery*, 59, 1519–1523.

200. Sivam GP, Lampe JW, Ulness B et al. (1997) Helicobacter pylori—in vitro susceptibility to garlic (Allium sativum) extract. *Nutrition and Cancer*, 27, 118–121.

201. Smythies JR. (1997a) The biochemical basis of synaptic plasticity and neurocomputation: a new theory. *Proceedings of the Royal Society of London. B*, 264, 575–579.

202. Smythies JR. (1997b) Oxidative reactions and schizophrenia: a review. *Schizophrenia Research*, 24, 357–364.

203. Smythies JR and Tolbert L. (1981) "Neuropharmacological roles for methionine, nicotinamide and ascorbic acid." In *Nutrition and Behavior* (SA Miller, ed.). Philadelphia: The Franklin Institute Press, 263–267.

204. Snodderly DM. (1995) Evidence for protection against age-related macular degeneration by carotenoids and antioxidant vitamins. *American Journal of Clinical Nutrition*, 62 (suppl.), 1448S–1461S.

205. Stähelin HB, Gey KF, Eicholzer M et al. (1991) beta-carotene and cancer prevention: the Basel study. *American Journal of Clinical Nutrition*, 53 (suppl.), S265–S269.

206. Stampfer MJ, Hennekens CH, Manson JE et al. (1993) Vitamin E consumption and the risk of coronary disease in women. *New England Journal of Medicine*, 328, 1444–1449.

207. Stampfer MJ and Rimm EB. (1995) Epidemiological evidence for vitamin E in prevention of cardiovascular disease. *American Journal of Clinical Nutrition*, 62 (suppl.), S1365–S1369.

208. Steiner M, Glantz M, and Lekos A. (1995) Vitamin E plus aspirin compared with aspirin alone in patients with transient ischemic attacks. *American Journal of Clinical Nutrition*, 62 (suppl.), S1381–S1384.

209. Stephens N. (1997) Editorial. Anti-oxidant therapy for ischaemic heart disease: where do we stand? *Lancet*, 349, 1710–1711.
210. Stephens NG, Parsons A, Schofield PM et al. (1996) Randomized control trial of vitamin E in patients with coronary disease: Cambridge Heart Antioxidant Study (CHAOS). *Lancet*, 347, 781–786.
211. Suboticanec K, Folnegovic-Smalc V, Korbar M et al. (1990) Vitamin C status in chronic schizophrenia. *Biological Psychiatry*, 28, 959–966.
212. Tang AM, Graham NMH, and Saah AJ. (1996) Effects of micronutritional intake on survival of human immunodeficiency virus type I infection. *American Journal of Epidemiology*, 143, 1244–1256.
213. Tavani A and La Vecchia C. (1995) Fruit and vegetable consumption and cancer risk in a Mediterranean population. *American Journal of Clinical Nutrition*, 61 (suppl.), 1374S–1377S.
214. Taylor A. (1992) Role of nutrients in delaying cataracts. *Annals of the New York Academy of Sciences*, 669, 111–123.
215. Tengerdy RP. (1990) The role of vitamin E in immune response and disease resistance. *Annals of the New York Academy of Sciences*, 587, 24–33.
216. Todd S, Woodward M, Bolton-Smith C et al. (1995) An investigation of the relationship between antioxidant vitamin intake and coronary heart disease in men and women using discriminant analysis. *Journal of Clinical Epidemiology*, 48, 297–305.
217. Toohey L, Harris MA, Allen KGD et al. (1995) Plasma ascorbic acid concentrations are related to cardiovascular risk factors in African-Americans. *Journal of Nutrition*, 126, 121–128.
218. Trichopoulou A. (1995) Editorial. Olive oil and breast cancer. *Cancer Causes and Control*, 6, 475–476.
219. Túri S, Németh I, Torkos A et al. (1997) Oxidative stress and antioxidant defence mechanism in glomerular diseases. *Free Radical Biology and Medicine*, 22, 161–168.
220. Urivetsky M, Kessaris D, and Smith AD. (1992) Ascorbic acid overdosing: a risk factor for calcium oxalate nephrolithiasis. *Journal of Urology*, 147, 1215–1218.
221. Van der Hagen AM, Yolton DP, Kaminski MS et al. (1993) Free radicals and anti-oxidant supplementation: a review of their roles in age-related macular degeneration. *Journal of the American Optometric Association*, 64, 871–878.
222. Van der Vliet A, Eiserich JP, Marelich GP et al. (1997) Oxidative stress in cystic fibrosis: does it occur and does it matter? *Advances in Pharmacology*, 38, 491–513.
223. Velussi M, Cernigoi AM, De Monte A et al. (1997) Long-term (12 months) treatment with an anti-oxidant drug (silymarin) is effective on hyperinsulinemia, exogenous insulin need and malondialdehyde levels in cirrhotic diabetic patients. *Journal of Hepatology*, 26, 871–879.
224. Voelker R. (1994) Recommendations for antioxidants: how much evidence is enough? *Journal of the American Medical Association*, 271, 1148–1149.
225. Voelker R. (1997) Antioxidants and asthma. *Journal of the American Medical Association*, 277, 1926.
226. Walker ARP. (1997) "Public nutrition": who is listening, responding, and acting. *Nutrition Research*, 17, 759–773.
227. Watanabe H, Kakihana M, Ohtsuka S et al. (1997) Randomized, double-blind, placebo-controlled study of supplemental vitamin E on attenuation of the development of nitrite tolerance. *Circulation*, 96, 2545–2550.
228. Weber P, Bendich A, and Machlin LJ. (1997) Vitamin E and human health: rationale for determining recommended intake levels. *Nutrition*, 13, 450–460.

229. Wechsler H, Levine S, Idelson RK et al. (1996) The physician's role in health promotions revisited—a survey of primary care practitioners. *New England Journal of Medicine*, 334, 996–998.

230. Weisburger JH. (1991) Nutritional approach to cancer prevention with emphasis on vitamins, antioxidants, and carotenoids. *American Journal of Clinical Nutrition*, 53 (suppl.), S226–S237.

231. Weisburger JH. (1995) Editorial. Vitamin C and disease prevention. *Journal of the American College of Nutrition*, 14, 109–111.

232. Winklhofer-Roob BM, Ellemunter H, Frühwirth M et al. (1997) Plasma vitamin C concentrations in patients with cystic fibrosis: evidence of associations with inflammation. *American Journal of Clinical Nutrition*, 65, 1858–1866.

233. Winyard PG and Blake DR. (1997) Antioxidants, redox-regulated transcription factors, and inflammation. *Advances in Pharmacology*, 38, 403–421.

234. Wiseman H and Halliwell B. (1996) Damage to DNA by reactive oxygen and nitrogen species: role in inflammatory disease and progression to cancer. *Biochemical Journal*, 313, 17–29.

235. Wiseman H, Plitzanopoulou P, and O'Reilly J. (1996) Antioxidant properties of ethanolic and aqueous extracts of green tea compared to black tea. *Biochemical Society Transactions*, 24 (suppl.), 390S.

236. Work Study Group on Diet, Nutrition, and Cancer of the American Cancer Society. (1991) Ca-a Cancer Journal for Physicians, 41, 334–338.

237. Yokoyama H, Lingle DM, Crestanello JA et al. (1996) Coenzyme Q_{10} protects coronary endothelial function from ischemia reperfusion injury via an antioxidant effect. *Surgery*, 120, 189–196.

238. Yong L-C, Brown CC, Schatzkin A et al. (1997) Intake of vitamins E, C, and A and risk of lung cancer: the NHANES I Epidemiological follow-up study. *American Journal of Epidemiology*, 146, 231–243.

239. Zhang H-M, Wakisaka N, Maeda O et al. (1997) Vitamin C inhibits the growth of a bacterial risk factor for gastric carcinoma: *Helicobacter Pylori*. *Cancer*, 80, 1897–1903.

240. Zheng W, Doyle TJ, Kushi LH et al. (1996) Tea consumption and cancer incidence in a prospective cohort study of postmenopausal women. *American Journal of Epidemiology*, 144, 175–182.

241. Ziegler D and Gries FA. (1997) α-lipoic acid in the treament of diabetic peripheral and cardiac autonomic neuropathy. *Diabetes*, 46 (suppl.), S62–S66.

242. Ziegler D, Schatz H, Conrad F et al. (1997) Effects of treatment with the antioxidant alpha-lipoic acid on cardiac autonomic neuropathy in NIDDM patients. *Diabetes Care*, 20, 369–373.

243. Ziegler RG. (1991) Vegetables, fruits, and carotenoids and the risk of cancer. *American Journal of Clinical Nutrition*, 53 (suppl.), S251–S259.

244. Ziegler RG, Mayne ST, and Swanson CA. (1996a) Nutrition and lung cancer. *Cancer Causes and Control*, 7, 157–177.

245. Ziegler RG, Colavito EA, Hartge P et al. (1996b) Importance of alpha-carotene, beta-carotene, and other phytochemicals in the etiology of lung cancer. *Journal of the National Cancer Society*, 88, 612–615.

246. Ziegler R, Nomura A, Craft N et al. (1997) Individual carotenoids in the etiology of lung and upper aerodigestive tract cancers. *American Journal of Epidemiology*, 145 (suppl.), S79.

John Smythies is a medical doctor and senior research fellow at the Institute of Neurology, Queen Square, London; director of the Division of Neurochemistry and Integrative Medicine, Center for Brain and Cognition, at the University of California, San Diego; and emeritus professor at the University of Alabama Medical Center, Birmingham. He has served as president of the International Society for Psychoneuro-endocrinology, editor of the International Review of Neurobiology from 1958 to 1991, and consultant to the World Health Organization. He is the author of more than two hundred scientific papers and thirteen books. He was the coauthor of the first specific biochemical theory of schizophrenia in 1952. His present research concentrates on the role of oxidative stress and antioxidants in the brain in health and in diseases such as schizophrenia and Parkinson's disease. Dr. Smythies also continually monitors and evaluates the flood of new information on oxidative stress and antioxidants published in the medical literature.

Dr. Smythies can also be reached at http://www-psy.ucsd.edu/~smythies

about the author